LOW TEMPERATURE PROCESSES

ENERGY SCIENCE, ENGINEERING AND TECHNOLOGY

Additional books in this series can be found on Nova's website
under the Series tab.

Additional E-books in this series can be found on Nova's website
under the E-book tab.

ENERGY SCIENCE, ENGINEERING AND TECHNOLOGY

LOW TEMPERATURE PROCESSES

MOUHAMADOU BASSIR DIOP

Nova Science Publishers, Inc.
New York

Copyright © 2012 by Nova Science Publishers, Inc.

All rights reserved. No part of this book may be reproduced, stored in a retrieval system or transmitted in any form or by any means: electronic, electrostatic, magnetic, tape, mechanical photocopying, recording or otherwise without the written permission of the Publisher.

For permission to use material from this book please contact us:
Telephone 631-231-7269; Fax 631-231-8175
Web Site: http://www.novapublishers.com

NOTICE TO THE READER
The Publisher has taken reasonable care in the preparation of this book, but makes no expressed or implied warranty of any kind and assumes no responsibility for any errors or omissions. No liability is assumed for incidental or consequential damages in connection with or arising out of information contained in this book. The Publisher shall not be liable for any special, consequential, or exemplary damages resulting, in whole or in part, from the readers' use of, or reliance upon, this material. Any parts of this book based on government reports are so indicated and copyright is claimed for those parts to the extent applicable to compilations of such works.

Independent verification should be sought for any data, advice or recommendations contained in this book. In addition, no responsibility is assumed by the publisher for any injury and/or damage to persons or property arising from any methods, products, instructions, ideas or otherwise contained in this publication.

This publication is designed to provide accurate and authoritative information with regard to the subject matter covered herein. It is sold with the clear understanding that the Publisher is not engaged in rendering legal or any other professional services. If legal or any other expert assistance is required, the services of a competent person should be sought. FROM A DECLARATION OF PARTICIPANTS JOINTLY ADOPTED BY A COMMITTEE OF THE AMERICAN BAR ASSOCIATION AND A COMMITTEE OF PUBLISHERS.

Additional color graphics may be available in the e-book version of this book.

Library of Congress Cataloging-in-Publication Data

Low temperature processes / editor, Mouhamadou Bassir Diop.
    p. cm.
Includes bibliographical references and index.
ISBN 978-1-62100-038-9 (hardcover)
1. Brickmaking--Energy conservation. 2. Cement. I. Diop, Mouhamadou Bassir, 1957-
TP827.L69 2011
666'.737--dc23
                            2011030122

Published by Nova Science Publishers, Inc. † *New York*

CONTENTS

PREFACE

The world is consuming more than 10 billion tons of materials every year. Much of this consumption is driven by production, use, and disposal of construction materials such as concrete, wood, steel, and brick. The cost, energy intensity, and environmental impact (e.g., CO_2 emission) of conventional building materials have created a significant impetus for exploring new techniques to produce materials with lower embodied energy. This book introduces a promising technology where new cementing materials are produced by alkali activation of natural or waste siliceous materials such as volcanic ash, clay, or coal fly ash. The resulting binder is an alumino-silicate phase forming an inorganic (geo-)polymer. A novel application of this method is introduced for manufacturing bricks at significantly lower temperatures ($\sim100^{\circ}$C) than the conventional sintering temperatures ($900\sim1000^{\circ}$C). In addition to major energy and cost savings, the method does not require the use of brick kilns and can be implemented in rural areas, solely by using the thermal energy of sunlight. The materials science principles behind the geo-polymerization process are elaborately discussed. In addition to bricks, this technology has the potential of broadly impacting the construction industry by producing a new generation of green Portland cement-free concrete materials. This monograph will catalyze new ideas and lead to new, innovative sustainable products.

Prof. Farshad Rajabipur
Energy and Mineral Engineering
Materials Research Institute
The Pennsylvania State University

Chapter 1

INTRODUCTION

Portland cement clinker is made by heating, in a kiln, a homogeneous mixture of raw materials to a sintering temperature, which is about 1450 °C for modern cements.

Portland cement manufacture can cause environmental impacts at all stages of the process. These include emissions of airborne pollution in the form of dust, gases, noise and vibration when operating machinery and during blasting in quarries, consumption of large quantities of fuel during manufacture, release of CO_2 from the raw materials during manufacture, and damage to countryside from quarrying.

Workers at Portland cement facilities, particularly those burning fuel containing sulfur, should be aware of the acute and chronic effects of exposure to SO_2 [sulfur dioxide]. Today, the most important environment, health and safety performance issues facing the cement industry are atmospheric releases (including greenhouse gas emissions, dioxin, NO_x, SO_2, and particulates), accidents and worker exposure to dust.

The thrust of innovation for the future is to reduce of CO_2 by modification of the chemistry of cement, by the use of wastes, and by adopting more efficient processes.

Portland cement has been discovered on 1874 in UK, rapidly its use have been generalized in construction for several reasons. First, it enables to have high resistance at early age. On the long term, the mechanical resistance increases continuously. If, its composition is appropriate (example pouzzolanic cement), it shows a good resistance to aggressive environment. Last, the major raw material usually limestone ($CaCO_3$) and the second raw materials (clay, shale, sand, iron ore etc) used are relatively abundant in the earth crust.

The manufacturing of bricks is an energy-intensive industry that contributes significantly to carbon dioxide emissions. Emissions from brick manufacturing facilities include particulate matter (MP), PM less than or equal to 10 μm in aerodynamic diameter (PM-10), PM less than or equal to 2.5 μm in aerodynamic diameter (PM-2.5), sulfur dioxide (SO_2), sulfur trioxide (SO_3), nitrogen oxide (NO_x), carbone monoxide (CO), carbone dioxide (CO_2), metals, total organic compounds (TOC), including methane, ethane, volatile organic compounds (VOC), and some hazardous air pollutants (HAP), hydrochloric acid (HCl) and fluoride compounds. More than 50 chemical pollutants are generated by the process. In 2002, over eight billion bricks were sold in the United States. Typically, bricks are made by firing clay to approximately 2,000 °F. Considering the fact that the manufacturing of one brick consumes around 2 Mega Joules/kg brick, the total energy used in USA can be estimated at sixteen billion Mega Joules/kg brick. The heating process changes the molecular structure of the clay in such a way that it is vitrified. Vitrification is not a necessary method to develop a brick process.

All techniques that use heating material around 1000°C generate toxic gases.

So the purpose of the book is to show the interest of new approaches to making cement in a broad sense that reduce energy by working at low temperature (< 100°C) and then eliminate pollutions. Cement should be understood as any substance that stick things together and resist to water effect. Such ideas seem "revolutionary" but they appear to be realistic because nature makes rocks at nearly ambient temperature. Since geological processes can take millions of years we have to use our intelligence and ingenuity to make things happen at human scale: a few weeks, days, or hours. Our recent work at Material Research Laboratory (MRI) at Penn State University proves beyond doubt that it's possible.

The interest of this approach is that it can be accomplished on a small or industrial scale as needed, on one hand, and use starting materials that can consist of natural materials as well as industrial or mining by product materials, on the other hand. This enables to solve environmental problems. Last the process does not cause pollution.

The purpose of this book is not to solve all the problems but to push all the researchers in building materials to develop and share new ideas, to think in a different way ... If researchers come to believe that it's possible the aim of this book will be attained an half of the way accomplished. I am convinced that it is the way that leads to sustainable development.....The good researcher is the one who has the good idea and the good process like Egyptians some thousand years ago...Portland cement making process is no more a good process or a good idea ....

Chapter 2

GEOLOGY

SECTION A: GENERAL GEOLOGY

Introduction

According to their genesis, the rocks, which are natural mineral aggregates, were divided into 3 categories:

1. The endogenous rocks called also igneous, eruptive, magmatic or plutonic, would be due in theory to the solidification of a melted silicated mass or magma. It is currently known that certain rocks of this category did not pass by a magmatic stage. Also the endogenous term must it be preferred.
2. Exogenic or sedimentary rocks coming from the accumulation of sediments cluster of mineral remains, organic or chemical deposits coming from the disintegration of pre-existing rocks.
3. Metamorphic rocks born from the transformation of endogenous or exogenic rocks. The distinction between certain endogenous rocks and certain very metamorphic rocks is often very difficult to establish.

Sedimentary rocks are the result of lithification of sediments, the loose material that accumulates at the surface as the result of weathering and transport and deposition of the weathered materials. Lithification is a rock forming process that involves:

- *Consolidation* - This is mostly due to pressure. The weight of overlying sediments results in compaction, a reduction in pore space and removal of water
- *Cementation* - The deposition from solution of a soluble substance. This fills any spaces between grains (pore space) and cements or glues the grains together. There are three common types of cement:
  - Calcite, probably the most common because calcium is easily dissolved in groundwater.
  - Silica. This is less soluble than calcite. It forms much stronger and harder cement.
  - Iron oxide (Fe_2O_3). This is the red mineral hematite and is the reason for the red color of some rocks.

Limestones

The igneous rocks contain on average calcium 3.6%. It enters in many silicates mineral composition indeed: plagioclases, garnets, epidotes, certain pyroxenes and amphiboles, and rocks the marnes, limestones and dolomites, etc Released by the processes of deterioration of the rocks, Ca enters in solution and water transport it in the form of ($CaH_2(CO_3)_2$) ionized (bicarbonate or acid carbonate Ca). It is brought by the rivers to the oceans. The differences between fresh water and sea water are very marked. Thus in the first the report/ratio Na /Ca is of 0.28, while in the seconds, it is of 26.63, that is to say 100 times higher. This calcium fall at sea is due to the behavior of Ca^{2+} in solution. A weak increase in the pH is enough to cause its precipitation in the form of $CaCO_3$. The current sediments would contain on average 12.5% of CaO, according to Kuenen (1956).

The annual contribution of the rivers is estimated by Clark at 6 billion tons of $CaCO_3$; this quantity is equal to that which settles each year on the sea-bed. That means that the sea water is saturated with $CaCO_3$.

$$CaCO_3 \rightleftharpoons Ca^{2+} + CO_3^{2-}$$

Dissociation is very weak, $CaCO_3$ is not very soluble. On the other hand, the bicarbonate is much more

$$CaH_2(CO_3)_2 \rightleftarrows Ca^{2+} + 2HCO_3$$

$$H^+ + CO_3^{2-}$$

This release of ions of weak acid (HCO_3) and strong cations (Ca^{2+}) gives to sea water a low alkalinity (pH~8). It is thus the presence of carbonic gas which maintains Ca^{2+} in solution. However its solubility varies with the temperature of water. It is much stronger in the cool water (double) that in warm water. The algae which absorb CO_2 and raise the pH contribute to the precipitation of $CaCO_3$. However, the development of the algae depends on the content of water phosphates and out of nitrates; these substances are especially abundant on the continental platform. Calcareous sedimentation will be thus especially the prerogative of warm water and not very deep.

Precipitated Limestones

In supersaturated warm water, the calcium carbonate can precipitate in certain circumstances, either on remains agitated by the waves, or within water in small letters needles of aragonite. The first mode gives rise to oolitic limestones, the second with lithographic limestones. Oolitic limestones are very widespread rocks, formed by the meeting of grain limestones, rounded and gauged well, assembled by clear calcite cement (sparrite). Each grain (ooïde, ovulite or oolite- this last name are also given to the rock) is consisted a core (remains of shell, foraminifère, grain of sand) surrounded by several envelopes of radiant calcite prisms. Oolites are formed currently at red sea, in the Bahamas, in the Big lake Salted of Utah etc always in the coastal hot and agitated water with depths lower than 3m. On the cores unceasingly turned over by the waves, $CaCO_3$ crystallizes in concentric plates of aragonite. Later, the unstable aragonite will be replaced by radiant calcite. Oolitic limestones are frequently formed in the zone of the back-reef.

The benches of the Bahamas are locally covered with a made mud of fine needles of aragonite. They are crystals generated within the water supersaturated by a fall of the content CO_2 due to the action of the unicellular algae. It is the phenomenon of the "whiting" observed in the Persian Gulf, where the local development of the plankton causes in the water large clear spots, kinds of crystal clouds of aragonite, which slowly elutriate and reach the bottom. Compacted, and calcite having replaced the aragonite, these vases will be transformed into the limestones with very fine grain, used in lithography, from where their lithographic limestone name. They do not have all this origin, some are calcilutites.

Tuffs and Trevertins

These rocks of beige, yellow or brown color, represent a zoned structure, due to successive encrustings of $CaCO_3$. They occur in the vicinity of certain sources, whose water is rich in calcium bicarbonate. At Griffon, part of CO_2 emerges, which involves the precipitation of $CaCO_3$. These relatively tender rocks are often used in the construction industry.

Siliceous Rock

Siliceous rock, any of a group of sedimentary rocks that consist largely or almost entirely of silicon dioxide (SiO_2), either as quartz or as amorphous silica and cristobalite; included are rocks that have formed as chemical precipitates and excluded are those of detrital or fragmental origin.

The most common siliceous rock is chert, which is a dense, microcrystalline rock composed of chalcedony and quartz. Chert is the second most abundant chemically precipitated rock after limestone. It occurs in beds and in nodules. Bedded chert consists of siliceous fossils such as diatoms and radiolarian.

Siliceous rocks are sedimentary rocks that have silica (SiO_2) as the principal constituent. The most common siliceous rock is Chert other types include Diatomite. They commonly form from silica-secreting organisms such as radiolarians, diatoms, or some types of sponges.

The chemical compound silicon dioxide, also known as silica (from the Latin *silex*), is an oxide of silicon with the chemical formula SiO_2. It has been known for its hardness since antiquity. Silica is most commonly found in nature as sand or quartz, as well as in the cell walls of diatoms.

Chert is a fine-grained silica-rich microcrystalline, cryptocrystalline or microfibrous sedimentary rock that may contain small fossils. It varies greatly in color (from white to black), but most often manifests as gray, brown, grayish brown and light green to rusty red; its color is an expression of trace elements present in the rock, and both red and green are most often related to traces of iron (in its oxidized and reduced forms respectively). Chert occurs as oval to irregular nodules in greensand, limestone, chalk, and dolostone formations as a replacement mineral, where it is formed as a result of some type of diagenesis. Where it occurs in chalk, it is usually called flint. It also occurs in thin beds, when it is a primary deposit (such as with many jaspers and radiolarites). Thick beds of chert occur in deep geosynclinal deposits. These thickly bedded cherts include the novaculite of the Ouachita Mountains of Arkansas, Oklahoma, and similar occurrences in Texas in the United States. The banded iron formations of Precambrian age are composed of alternating layers of chert and iron oxides. Chert also occurs in diatomaceous deposits and is known as diatomaceous chert. Diatomaceous chert consists of beds and lenses of diatomite which were converted during diagenesis into dense, hard chert. Beds of marine diatomaceous chert comprising strata several hundred meters thick have been reported from sedimentary sequences such as the Miocene Monterey Formation of California and occur in rocks as old as the Cretaceous. Diatomaceous earth also known as diatomite or kieselgur, is a naturally occurring, soft, siliceous sedimentary rock that is easily crumbled into a fine white to off-white powder. It has a particle size ranging from less than 1 micrometre to more than 1 millimeter, but typically 10 to 200 micrometres. This powder has an abrasive feel, similar to pumice powder, and is very light, due to its high porosity. The typical chemical composition of oven dried diatomaceous earth is 80 to 90% silica, with 2 to 4% alumina (attributed mostly to clay minerals) and 0.5 to 2% iron oxide. Diatomaceous earth consists of fossilized remains of diatoms, a type of hard-shelled algae.

Silica and Water

The silica which is formed by reaction from solutions at P_{atm} presents in the form of gel which evolves slowly while being dehydrated to the stable crystallized phases.

$$Si(OH)_4 \longrightarrow H^+ + SiO(OH)_3 - \text{followed by}$$

$$SiO(OH)_3 \longrightarrow H^+ + SiO_2(OH)_2{}^{2-}$$

The action of water on silica at ordinary temperature is complex. It is found, in solution $Si(OH)_4$ which is polymerized easily to give solid gel and polymers. One recognized for silicic acid an ionization in 2 stages:

$$Si(OH)_4 \longrightarrow H^+ + SiO(OH)_3 \text{ follow-up by}$$

$$SiO(OH)_3 \longrightarrow H^+ + SiO_2(OH)_2{}^{2-}$$

$SiO_2 (OH)_2^{2-}$ is an acid weaker than the carbon dioxide with a very precarious stability in very diluted solution. Oligosilicic acids exist also, at least one them is known by its esters, it is the disilicic acid obtained from the precedent by dehydration:

$(OH)_3$ Si-O - Si $(OH)_3$ which could continue and give $(OH)_3$ Si-O - Si $(OH)_2$ - O - Si $(OH)_3$ reaction which seems reversible by hydrolyzing cut…

One could continue but as soon as trisilicic acid is obtained desiccant condensation can be done on an OH group of central Si and the molecule ceases then to be linear. One can lead to polydimensional macromolecules and by construction of a network, more or less coherent which binds the molecules between them, one obtains a gel (very hygrophile: the silicagels are well-known as desiccants) in which pore water is saturated with amorphous silica. One can drive out water of these compounds by increasing the pressure and one descends thus; to 4 to 6% of water fixed by inclusion, adsorption or H connection. It is then necessary to increase the temperature but even in the synthesis of quartz thus obtained, the Raman spectrum indicates the presence of some residual OH groups. Silica behaves in water like an anhydride and in dissolving inside it, water becomes a slightly conducting liquid of pH 4.7 at saturation. That is to say for a concentration in ion H^+ of 2/100000 or a molar concentration of 215/100000, one can admit that the solid reacts with water. The solubility of amorphous silica in water is very weak from 0.05 to 0.08 g/l at 0°C and till 0.44 g/l at 100°C. It depends very little on the pH till it reaches 9. But if the pH exceeds 9 the molecule $Si(OH)_4$ ionize itself and solubility increases then strongly: it reaches 4000 ppm with pH 11, for example (case of certain evaporitic lakes). But all this is true; only with a constant ion nature. Indeed, the solubility of silica is not function of the pH alone, but also of the involved ions. So Fe^{3+} in acid solutions (pH 1.5 to 3) produced a dissociation of silica much more important than Ca^{++} or NH^{4+} whose solutions have pH slightly acid, neutral, and even alkaline. Among the other ions, alone Al^{+++} and Mg^{++} affect solubility in lowering it. It will form a thin protective coating of aluminium or of magnesium silicate … Quartz is less soluble than amorphous silica and for the powders the solubility also depends on the diameter of the grains. The solubility of quartz or opal grains whose diameters are higher than 250 μm is always practically nul after 200 days at 20°C. If the grains measure 5 μm and were removed from an amorphous layer by crushing, one obtains a solubility of 0.007 g/l at 20°C. But for the earth diatoms with opal, more than 200 days are necessary, in all the cases, to reach this value.

Organogenic Siliceous Rocks

The organogenic siliceous rocks are primarily made up by silica (quartz, chalcedony, opal) coming from organisms with siliceous test.

The decomposition of silicates, particularly in the tropical areas, is accompanied by the solubilization of silica in the form of ortho silicic acid H_4SiO_4. The solubility of this acid between 120 and 140 PPM is independent of the pH met under natural conditions.

The rivers transport this dissolved silica with contents varying between 10 and 60 PPM. Despite of this regular contribution, marine water is low in silica (0.1 to 5 PPM). The difference comes from its consumption by the following organisms:

Radiolarian (actinopodes unicellular)
Silicoflagellés (unicellular whipped)
Ebriédiens "
Diatoms (unicellular algae)
Siliceous sponges (métazoaires)

Phosphates

Phosphorus is an additional element of the igneous rocks. They contain on average 1.2 kg per ton, primarily in the form of apatite Ca (OH, F, CI) $(PO_4)_3$. It crystallizes in small hexagonal needles when the first minerals form.

During the weathering of the rocks which contain some, apatite is solubilized in the form of alkaline phosphate or acid calcium phosphate. A part is drawn from the ground by the plants (phosphorus is an element necessary of the cytoplasm), the remainder is pulled by subterranean water and is poured in the rivers.

They contain an average between 5 to 10 mg/m^3 of apatite. In marine water, the phosphorus content is low in the photic zone (action of the algae). The deep water on the other hand is saturated.

The marine animals which introduce the algae assimilate part of this phosphate. Some of incorporate it in their skeleton: the inarticulate brachiopods whose shell can titrate till 75% of Ca_3 P_2O_5 and especially the fish, of which the skeleton, like that of all the vertebrate ones, is made of calcium phosphate.

Ferrous Rock

Nearly all of Earth's major iron ore deposits are in rocks that formed over 1.8 billion years ago. At that time Earth's oceans contained abundant dissolved iron and almost no dissolved oxygen. The iron ore deposits began forming when the first organisms capable of photosynthesis began releasing oxygen into the waters. This oxygen immediately combined with the abundant dissolved iron to produce hematite or magnetite. These minerals deposited on the sea floor in great abundance, forming what are now known as the "banded iron formations." The rocks are "banded" because the iron minerals deposited in alternating bands with silica and sometimes shale. The banding might have resulted from seasonal changes in organism activity.

Magnetite is a ferrimagnetic mineral with chemical formula Fe_3O_4, one of several iron oxides and a member of the spinel group. The chemical IUPAC name is iron(II,III) oxide and the common chemical name is ferrous-ferric oxide. The formula for magnetite may also be written as $FeO \cdot Fe_2O_3$, which is one part wüstite (FeO) and one part hematite (Fe_2O_3). This refers to the different oxidation states of the iron in one structure, not a solid solution. The Curie temperature of magnetite is 858 K (585°C; 1,085°F). It is black or brownish-black with a metallic luster, has a Mohs hardness of 5–6 and a black streak.

Hematite, also spelled as haematite, is the mineral form of iron (III) oxide (Fe_2O_3), one of several iron oxides. Hematite crystallizes in the rhombohedral system, and it has the same crystal structure as ilmenite and corundum. Hematite and ilmenite form a complete solid solution at temperatures above 950°C.

Hematite is a mineral, colored black to steel or silver-gray, brown to reddish brown, or red. It is mined as the main ore of iron. Varieties include *kidney ore*, *martite* (pseudomorphs after magnetite), *iron rose* and *specularite* (specular hematite). While the forms of hematite vary, they all have a rust-red streak. Hematite is harder than pure iron, but much more brittle. Maghemite is a hematite- and magnetite-related oxide mineral.

Huge deposits of hematite are found in banded iron formations. Grey hematite is typically found in places where there has been standing water or mineral hot springs, such as those in Yellowstone National Park in the United States. The mineral can precipitate out of water and collect in layers at the bottom of a lake, spring, or other standing water. Hematite can also occur without water, however, usually as the result of volcanic activity.

Clay-sized hematite crystals can also occur as a secondary mineral formed by weathering processes in soil, and along with other iron oxides or oxyhydroxides such as goethite, is responsible for the red color of many tropical, ancient, or otherwise highly weathered soils.

Conclusion

Last we know that most of sedimentary rocks (carbonates, ferrous, siliceous, phosphates rocks etc...) are forming at ambient temperature at the surface of the earth crust or in the ocean. But they happen at geological scale (millions of years). To use these processes and make them happen at human scale, we have to work some parameters causing these processes: concentration, pH, alkalinity, composition of raw material, temperature, pressure etc...

References

H Badoux, 1989, Course of General Geology,

Kuenen, Ph. H. (1956): Classification of carbonate rocks. - American Association of Petroleum Geologists, Journal of Sedimentary Petrology, 22, 64-72...

Sam Boggs, Jr., "Principles of Sedimentology and Stratigraphy", Prentice Hall, 2006, 4th Ed.,

George R. Rapp, "Archaeomineralogy", 2002.

Barbara E. Luedtke, "The Identification of Sources of Chert Artifacts", American Antiquity, Vol. 44, No.4 (Oct., 1979), 744-757.

W.L. Roberts, T.J. Campbell, G.R. Rapp Jr., "Encyclopedia of Mineralogy, Second Edition", 1990

R.S. Mitchell, "Dictionary of Rocks", 1985.

SECTION B: GEOLOGY OF SENEGAL

GENERAL PRESENTATION
Official name: Republic of Senegal
Capital city: Dakar

Total population (July 2000 estimate): 9,987,000
Area: 196,722 sq.km, (76,124 sq.miles)
Annual population growth rate (2000): 2.94%
Life expectancy at birth (1998): 52.7 years
People not expected to survive to age 40 (1998): 28% of total population
GDP per capita (2009): 1,066 US $,
Currency: CFA Franc (XOF)
1 Euro = 656 CFA Francs
Official language: French
Main religion: Islam
National holiday: April 4th

Senegal is located at the western-most part of Africa at the Atlantic Ocean. The Senegal, Gambia and Casamance Rivers drain from extensive inland plains with altitudes less than 200 m. In the south-east of the country, plateaux with altitudes up to 600 m form the foothills of the north-south striking Bassaride mountain range. North of the Gambia River, much of the land is barren except for the floodplains of the Senegal River.

The agricultural sector, dominated by crop production and coastal fishing, as well as the tourist and mining sectors, forms an important part of the Senegalese economy. In 1999, agriculture accounted for 18% of the GDP, and employed more than 60% of the working population. The staple food of the Senegalese is rice, followed by millet and sorghum. The main export crops of Senegal are groundnuts and cotton. Extensive sugarcane production provides a large part of the total national sugar requirement. The production and export of phosphate rock, and phosphate-based fertilizers dominate the mineral industry of Senegal and phosphate production has been relatively stable over the last few decades. Small occurrences of gold and industrial minerals are also reported from Senegal. Exploration for hydrocarbons has revealed limited offshore oil resources and substantial amounts of onshore natural gas (106 billion cubic feet).

Geological Outline

The substratum of the Senegalese territory is made up of two major geological domains: the shallow-dipping Upper Cretaceous to Quaternary sediments in most of the central and western parts of Senegal, which occupies more than 75% of the territory, and the Precambrian basement, and in the east by the Palaeoproterozoic volcano-sedimentary sequences of the Kedougou-Kenieba inlier.

Senegal is dominated by two major geological units: the folded Precambrian basement in the east of the country, and the shallow-dipping Upper Cretaceous to Quaternary sediments in most of the central and western parts of Senegal.

The Precambrian in the east and southeast of the country is subdivided into the Paleoproterozoic Birimian volcano-sedimentary sequence, the Neoproterozoic Madina-Kouta Basin Series, and the two folded Neoproterozoic/Cambrian Pan-African mountainous ranges, the Bassaride Branch and the Koulonton Branch. The Lower Cambrian is represented in the Faleme Basin with tillites, cherts and limestones. Between the two Neoproterozoic/Cambrian Pan-African sequences lies a basin filled with Cambro-Ordovician conglomerates, mudstones and sandstones.

Precambrian Basement

The Precambrian basement formations are constituted at the west by the Mauritanides range bordering the eastern part of the Sedimentary Basin and in the east by the Palaeoproterozoic volcano-sedimentary sequences of the Kedougou- Kenieba inlier.

The formations of the Mauritanides chain are Herycian age and constitute one of the mobile areas of the West African craton. They are known for their numerous copper and chromium occurrences which, in Mauritania, constitute the important copper deposits of the Akjout Region. The Palaeoproterozoic volcano-sedimentary sequences, mostly known as Birimian formations, are of great metallogenic importance, as far as they contain the major ore deposits discovered in the region. The Kedougou-Kenieba inlier is limited to the west by the Mauritanides chain, and on all other sides by the Upper Proterozoic and Cambrian sediments of the Basin of Taoudenni. The Kedougou-Kenieba inlier is interpreted as an accretion of north-easterly trending Birimian age volcanic terrains. It includes two major geological structures, the Senegalomalian Fault and the Main Transcurrent Zone (MTZ) to which gold mineralisation is associated. Recent combination of geological studies including field work, and structural modelling, and of detailed core logging have improved the understanding of the geological structure of the MTZ. Two main zones of mineralisation have been further refined based on the latest geological model. Geological studies suggest that mineralisation in the prospective Sabodala volcano sedimentary belt and the Senegal-Malian shear zone is associated with an altered and sulphidised gabbro, which has intruded along the main structure, and a typical shear zone, hosted, where a structure has developed at the contact between a package of volcaniclastics and sediments. A lapilli tuff acts as a prominent marker horizon in the hanging wall of mineralisation.

The inlier is divided into three main stratigraphic units from west to east: the Mako Supergroup, the Diale Supergroup and the Daléma Supergroup.

- The Mako supergroup hosts Sabodala deposits located in an area of intense shearing and silicification associated with pyrite gold mineralisation. It forms a north-east tectonic structure, turning to north-west near the border with Mali, in the north. Typical lithologies include basalt flows; often carbonate alterations and minor volcaniclastic intercalations, magnesium basalt or komatiites, ultramafic sub-volcanic intrusions (pyroxenites) and numerous massive biotite and amphibole granitoids. These granitoid intrusions are suspected to have been 'heat engines' which sparked off the deep mineralised magmatic fluids related to the belated mineralisation in the Kédougou-Kéniéba inlier.
- The Diale Supergroup, located between the Mako Supergroup and the western edge of the Saraya granite is weakly metamorphic. It includes extensively folded formations, deposited after those of the Mako Supergroup and consisting of shale, greywacke, quartzite and volcanodétritic rocks.
- The Dalema Supergroup, located between the Saraya granite and the Faleme River, continues to Mali in its eastern part but disappears in the South under the Segou Madina Kouta series. It is composed of volcano-sedimentary schist and grauwacke rocks.

These Birimian formations are affected by syn, late and post-tectonic granite intrusions. The Precambrian basement is a metallogenic province of major importance for Senegal, which hosts numerous deposits and anomalies of gold, iron, uranium, lithium, tin, molybdenum and nickel in Birimian formations, and copper and chromium in the Mauritanides range. In addition to these metal resources, there are large marble and other ornamental rocks deposits, but also non metallic indices and deposits of barytes, kaolin, asbestos etc.

Sedimentary Basin

The Senegal Basin occupies (Figure 1) the central part of the Northwest African Coastal Basin (MSGBC Basin), which extends from the Reguibat ridge at the north end of the Guinean fault. It is typical passive margin opening westward to the Atlantic Ocean and whose eastern limit is represented by the Mauritanides chains.

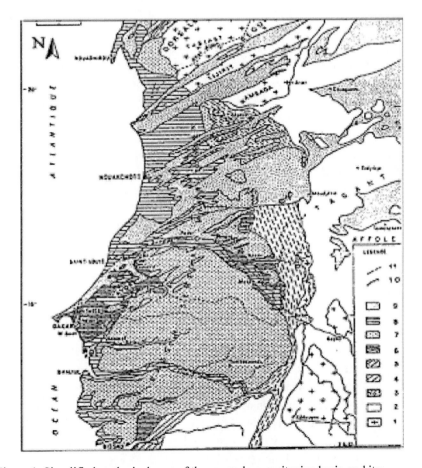

Figure 1. Simplified geological map of the senegalo-mauritanian basin and its surroundings. 1: Granitised Precambrien basement; 2: Upper Proterozoïc and Paleozoïc Sédiments; 3: Faulted mauritanides Chains; 4: Mesozoïc; 5: Weathered Mesozoic; 6: Paleogen; 7: weathered cenozoïc "continentale terminale"; 8: marine, lacustrin and alluvial quaternary sediments; 9: quaternary sediments (ancient and actual erg); 10 thrust front; 11 horst of Ndiass fault (after D.Nahon, 1976, modified).

The Senegal Sedimentary Basin is a Mesozoic Basin. It has gone through a complex history in relation to the pre-rift (Upper Proterozoic to Paleozoic), the Syn-rift (Permian to Triassic) and the Post-rift (Central Jurassic to Holocene) at different stages of development of the Basin.

Most of the outcrops of the basin are composed of recent sandy covers. Maestrichian and Eocene formations outcrop, however, in the peninsula of Cape Verde while Eocene outcrop in the valley of Senegal River. The description and knowledge of the Basin have been made possible largely thanks to hydraulic and oil drilling data.

The Secondary formations include Palaeocene zoogenic limestone exploited at Bandia and Pout by cement plants and aggregates producers. They include also Maestrichian sands, clays and sandstones. Paleocene and Maestrichian formations are also known to be major aquifers that contribute significantly to the water supply for cities and villages in the basin.

Tertiary formations hold into the Eocene compartment, significant resources of phosphates, limestone, attapulgite, clay and ceramics, solid fuels, etc. A major part of the basin is covered with superficial Quaternary formations, which in the middle and recent parts are characterised by fixed red sand dunes, semi-fixed or alive yellow and white dunes. These dunes, often exploited as building materials around urban centres, constitute also important reservoirs of heavy minerals.

Phosphates

There are several phosphate occurrences and deposits in Senegal (Figure 2). The four main phosphate deposits are:

- the Neoproterozoic/Cambrian phosphates in the Namel area, southeast Senegal,
- the Eocene phosphate deposits along the Senegal River, including the 'Matam' deposits,
- the Eocene primary phosphate deposits in western Senegal, mined at Taiba and Lam Lam, •
- the aluminous phosphates of Thies, weathering products of the Eocene phosphates, found also in western Senegal.

Senegal is one of the major phosphate producers in sub-Saharan Africa. In 1997, the total phosphate production from Senegal was 616,700 tonnes (British Geological Survey 1999), down from approximately 1 million tonnes of phosphate concentrate in 1993, which were exported to Canada, Australia, Mexico and China. A high proportion of the concentrate is used for the industrial processing and production of soluble P-fertilizers, for instance SSP, TSP, DAP and NPKs. Most of the processed P-fertilizers are exported.

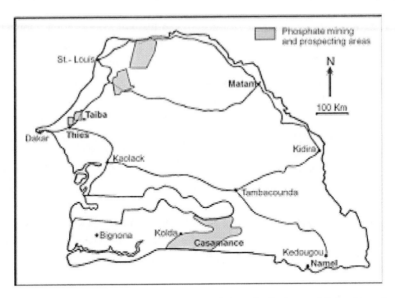

Figure 2. Location of phosphate occurrences and deposits in Senegal (after Pascal and Sustrac, 1989).

The phosphates of Taiba

The phosphate deposit of Taiba was discovered in 1948 and up to now has been the main phosphate mining area of Senegal. The deposit is mined by the Compagnie Sénégalaise des Phosphates de Taiba (CSPT). The phosphate beds are part of a very extensive phosphate-bearing area northeast of Dakar. Phosphate mining takes place mainly in the Keur Mor Fall deposit area at Taiba, some 110 km by rail from Dakar. Here, the Middle Eocene (Lutetian) phosphatic sequence can be divided into three major beds. They are, from top to bottom:

- 2-3 m homogenous phosphates,
- heterogeneous phosphatic ore containing flint,
- thin indurated coprolitic phosphates (phosphate gravel), 3-4 m thick.

Below the 5-12 m (average 7 m) thick phosphatic beds are Middle Eocene finely laminated 'paper' clays, made up largely of palygorskite (attapulgite). Above the phosphates are up to 25 m thick Quaternary aeolian sands (Pascal and Sustrac, 1989). The proven reserves of the Keur Mor Fall deposit are 100 million tonnes with ore grades ranging from 18-39% P_2O_5 (McClellan and Notholt, 1986). The average grade is 24% P_2O_5.

The neutral ammonium citrate solubility of the Taiba phosphate concentrate (37.4% P_2O_5) is 3.1% P_2O_5 (McClellan and Notholt 1986). The Cd contents of the Taiba phosphates are elevated, ranging from 60-115 mg/kg, and averaging 87 mg/kg. The aluminous phosphates of Thies are aluminous phosphatic rocks, resulting from long periods of weathering of phosphatic sediments, cover large parts of the Thies Plateau of western Senegal. The city of Thies is located in the centre of this extensive, elevated area. Natural outcrops of the aluminous phosphates are sparse. The best exposures are seen in open pits between Lam Lam and Pallo, approximately 15 km northwest of Thies. The aluminous phosphates of Thies are the result of lateritic weathering of the underlying Middle Eocene to Oligocene argillaceous phosphatic sediments. The weathering episode is estimated to have occurred from Middle Miocene to Lower Pliocene. For his doctorate thesis, Flicoteaux (1982) studied the genesis of this deposit in detail and found at least four stages of weathering: apatite leaching (stage 1), accumulation of kaolinite and Fe-millisite (stage 2), 'ochreous' aluminous phosphate development (stage 3), and leaching into 'white facies' phosphates in topographic depressions (stage 4). The main phosphatic weathering products are Ca-millisite, Sr-crandallite and wavellite. Mineralogical studies showed that the neutral ammonium citrate solubility of the typical Al-phosphate product of Pallo (32.0% P_2O_5) is high at 12.0% P_2O_5 (McClellan and Notholt 1986). The Société Sénégalaise des Phosphates de Taiba (SSPT) mines these aluminous phosphates in open-pit operations near the village of Pallo, 10 km northwest of Thies. The mineable phosphate ore at Pallo is 10 m thick and has an overburden of 3 m. The aluminous phosphates are crushed, calcined to increase the grade to 34% P_2O_5 and also to increase citrate solubility, and marketed as 'Phosal' for use as a fertilizer or 'Polyphos' used in animal feed (Flicoteaux and Hameh 1989). Proven aluminous phosphate reserves in the 32,000 hectare concession amount to 50 million tones with 29% P_2O_5 (Flicoteaux and Hameh 1989). Between 1979 and 1983 the annual production of crude phosphate ore from the aluminous phosphate deposit of Pallo was 180,000-280,000 tonnes, and the corresponding calcined ore between 78,000 and 140,000 tonnes (Flicoteaux and Hameh, 1989).

The Lam Lam Phosphate Deposit

Unweathered Ca-phosphates are mined by the Société Sénégalaise des Phosphates de Taiba (SSPT) near the village of Lam Lam, northwest of Thies. These unweathered phosphates resemble the phosphates of Taiba and occur as 7 m thick layers under a thick iron crust. Proven reserves of this deposit are 4 million tones of marketable product with an average grade of 33% P_2O_5. Only 1.5 million tones of these reserves have an overburden of less than 24 m (McClellan and Notholt, 1986).

The Matam (Ouali-Diala) Phosphate Deposits
The Matam phosphate deposits, described in detail by Pascal and Cheikh Faye (1989), occur on the left bank of the Senegal River. The phosphate beds can be traced over a distance of at least 100 km (Figure 3). The phosphatic layers are 4-16 m thick.

At places they lie at shallow depth along the slope of the Senegal River. In other places they occur more than 30 m below the flat and monotonous landscape. The first systematic prospecting took place between 1962 and 1966, and further work between 1980 and 1984 resulted in the delineation of this extensive phosphate deposit (Pascal and Faye 1989).

Palaeontological studies on shark teeth assign the Matam phosphates to the Lower Eocene, but some reworked coarse-grained coprolitic phosphates also occur in the Quaternary. The main phosphate beds in the N'Diendouri and Ouali Diala area are slightly indurated, light grey, coarse grained 'arenites.' Here, the phosphatic sequence is exposed at or near to the surface. The main phosphatic unit is 6-10 m thick in a clay-rich matrix containing mainly palygorskite (attapulgite) and montmorillonite.

The phosphate mineral has been identified as francolite. The neutral ammonium citrate solubility of the Matam phosphates (28.7% P_2O_5) is relatively high, 4.5% P_2O_5 (McClellan and Notholt, 1986) indicating that the chemical and mineralogical composition of the phosphate rocks is favourable for direct application in agriculture (Pascal and Faye, 1989).

The trace element content of this phosphate is low, with Cd concentrations of less than 5 mg/kg, and U concentrations of less than 40 mg/kg. Pascal and Faye (1989) reported seven major occurrences on the left bank of the Senegal River, two of them regarded as major deposits with reserves of more than 10 million tones. The overall reserves of the phosphates between N'Diendouri and Ouali Diala exceed 40 million tonnes at a grade averaging 28.7% P_2O_5. The deposit could be worked by open pit methods with a mean overburden-to-ore ratio of 4.5:1 (Pascal and Sustrac, 1989). Industries Chimiques du Sénégal (ICS) plans a new US $100 million phosphate mine in the Matam area (Leaky and Harrison, 2000).

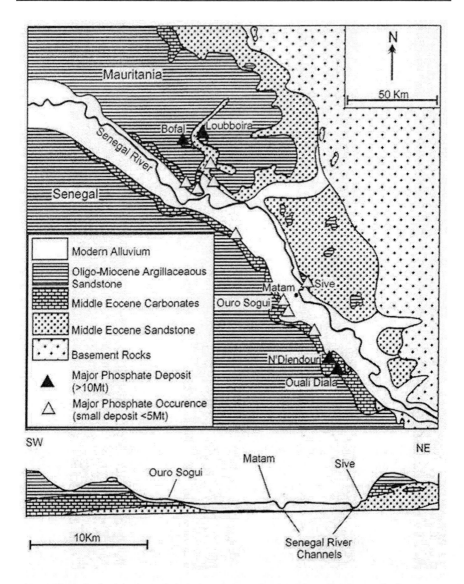

Figure 3. Geological setting of phosphate mineralization along the Senegal River near Matam, Senegal (Pascal and Faye, 1989).

The Namel Phosphate Deposit

The phosphate deposit near Namel in southeast Senegal was discovered during a systematic geochemical and geophysical exploration campaign in 1975. The Neoproterozoic to Lower Cambrian (approximately 650 million years of age) phosphates occur at the western edge of the Precambrian basement zone at the Kedougou inlier in eastern Senegal/northern Guinea (Pascal and Sustrac, 1989). The rock sequence unconformably overlies the Paleoproterozoic (Birimian) basement. Below the phosphate sequence lies a 3 m thick tillite bed. It is overlain by 18 m of fine-grained siliceous rocks, 9 m of phosphatic pelites, 2 m of calcareous phosphates (15% P_2O_5) and 8 m coarse phosphates (22-32% P_2O_5). The phosphate layers with over 20% P_2O_5 contain abundant silica (Slansky, 1986). The phosphate grains vary in size from 50 µm to several mm. The CaO/P_2O_5 ratios vary from 1.2 in the upper weathered zone to 1.8 at depth where carbonate gangue is present. Mudstones and dolomites overlie the phosphates. Mineralogical studies of phosphate samples collected from the surface show deformed grains of recrystallized fine-grained apatite in a siliceous matrix (McClellan and Saavedra, 1986). Crystallographic data indicate a nearly pure fluor-apatite (unit-cell a-value = 0.9373 Å). Banded calcareous phosphorites samples from 8 m depth show that the apatite is a low carbonate-substituted francolite with a crystallographic unit-cell a-value of 0.9360 Å (McClellan and Saavedra, 1986).

The best exposure of this phosphate sequence is in the vicinity of Namel. Here, the phosphates are exposed over a length of 1 km on the east side of the Namel Valley. No reserve data are available.

Other Phosphate Resources
There are several more Eocene phosphate 'occurrences' reported in Senegal, for instance near Thies, Sebikotane, Pointe Sarene and southeast of M'Bour (McClellan and Notholt, 1986). Notholt (1994) reported 20 million tonnes of phosphate resources from 'wastes' in the western part of Senegal. Material that is finer than 40 µm is currently discarded as 'waste' although it contains 26% P_2O_5. No more details on this potential resource are given.

Current Mining Operations
There are several phosphate companies operating in Senegal.
Compagnie Sénégalaise des Phosphates de Taiba (CSPT) mines and concentrates the Ca-phosphates at Taiba. The Senegalese government is a 50% shareholder of this company, which employed 1,400 persons in 1994. The mine and concentration-drying plant is located at Taiba. The concentrate is shipped by rail to the Dakar stockyard from where it is shipped abroad.
Société Sénégalaise des Phosphates de Thies (SSPT) mines mainly palygorskite (attapulgite) and minor amounts of Al-phosphate from Thies. In

1994, the government and the French company Rhone Poulence were shareholders and employed 230 persons. Industries Chimique du Sénégal (ICS) manufactures phosphate fertilizers from phosphate concentrates supplied by CSPT. ICS is located next to the CSPT mine site. Sulphuric acid and phosphoric acid plants are located at Darou Khoudoss and the fertilizer unit is located at Mbao. Shipping of the various fertilizers, including NPK, DAP, TSP and SSP, is from the free port of Dakar. ICS plans a major extension of its mine capacity and chemical facilities (Leaky and Harrison, 2000).

Limestones

Small occurrences of dolomitic limestones and marbles are known from the Paleoproterozoic in the Kedougou inlier in southeast Senegal, close to the border with Guinea. This greyish marble, 25 km west of Kedougou, has been exploited for the production of lime in the past (Roth 1996). Upper Cretaceous to Paleocene marly limestones occur east of Dakar in the Popenguine-Thies

area. The Paleocene limestones are massive and coarse-grained and are up to 40 m thick. Eocene limestones occur in several locations in Senegal, east of Dakar in the area of Rufisque, and along the banks of the Senegal River (Roth, 1996).

Other Agrominerals

Peat resources, estimated at 52 million m^3 occur along the coast of Senegal. A surveying permit was granted to Cie des Tourbières du Sénégal (Mining Annual Review, 1994).

Phospho-Gypsum

Large amounts of phospho-gypsum are produced annually by Industries Chimique du Senegal (ICS). Parts of these 'waste' products are currently being applied on Senegalese soils within the framework of the national program aiming at increasing agricultural production in Senegal. Long-term experiments with phospho-gypsum were started in 1997 and are ongoing (Sene et al. unpublished). In 1999, the utilization of phospho-gypsum in Senegal was 62,153 tones (Diop, pers. comm. Oct. 2001).

Agromineral Potential

The agromineral potential of the locally available phosphate resources is high not only for export but also for domestic use. Many small sedimentary phosphate beds occur in western Senegal, mostly, however, under considerable overburden. Extensive phosphatic sequences occur along the Senegal River and in the southeast of the country. Of considerable interest are

References

British Geological Survey 1999. World Mineral Statistics 1993-1997 - Production, exports, imports. Keyworth, Nottingham, UK:286p.

Burnett WC, Schultz MK and DH Carter 1996. Radionuclide flow during the conversion of phosphogypsum to ammonium sulfate. *J. Environm. Radioactivity,* 32:1-2, 33-51.

Diop AK 1999. Sustainable agriculture: New paradigms and old practices? Increased production with management of organic inputs in Senegal. *Env. Developm. Sustainability,* 1:285-296.

Flicoteaux R 1982. Genèse des phosphates alumineux du Sénégal occidental. Etapes et guides de l'altération. (Th. Sci. Univ. Marseille St-Jerome 1980). *Mem. Sci. Geol.* Univ. Strasbourg 67, 229p.

Flicoteaux R and PM Hameh 1989. The aluminous phosphate deposits of Thies, western Senegal. In: Notholt AJG, Sheldon RP and DF Davidson (eds.) Phosphate deposits of the world. Vol 2. Phosphate rock resources, Cambridge University Press, Cambridge, UK:273-276.

Leaky K and A Harrison 2000. Senegal. Mining Annual Review 2000. *The Mining Journal* Ltd. London.

Ledru P, Pons P, Milesi JP, Feybesse JL and V Johan 1991. Transcurrent tectonics and polycyclic evolution in the Lower Proterozoic of Senegal-Mali. *Precambr. Res.* 50:337-354.

McClellan GH and AJG Notholt 1986. Phosphate deposits of sub-Saharan Africa. In: Mokwunye AU and PLG Vlek (eds.) Management of nitrogen and phosphorus fertilizers in sub-Saharan Africa. Martinus Nijhoff, Dordrecht, Netherlands:173-224.

McClellan GH and FN Saavedra 1986. Proterozoic and Cambrian phosphorites - specialist studies: chemical and mineral characteristics of some Precambrian phosphorites. In Cook PJ and JH Shergold (eds.) Phosphate deposits of the world. Vol. 1. Proterozoic and Cambrian phosphorites. Cambridge University Press, Cambridge, UK:244-267.

Ministry of Energy, Mines and Industry of Senegal 1996. Senegal. In: *Mining Annual Review* 1996, p.165.

Notholt AJG 1994. Phosphate rock in developing countries. In: Mathers SJ and AJG Notholt (eds.) Industrial minerals in developing countries. AGID Rep. *Series Geosciences in International Development* 18:193-222.

Pascal M and M Cheikh Faye 1989. The Matam phosphate deposits. In: Notholt AJG, Sheldon RP and DF Davidson (eds.) Phosphate deposits of the world. Vol 2. Phosphate rock resources, Cambridge University Press, Cambridge, UK:295-300.

Pascal M and G Sustrac 1989. Phosphorite deposits of Senegal. In: Notholt AJG, Sheldon RP and DF Davidson (eds.) Phosphate deposits of the world. Vol 2. Phosphate rock resources, Cambridge University Press, Cambridge, UK:233-246.

Roth W 1996. Senegal. In: Bosse H-R, Gwosdz W, Lorenz W, Markwich, Roth W and F Wolff (eds.) Limestone and dolomite resources of Africa. *Geol. J.b.*, D, 339-344.

Sene M, Diack M and A Badiane (1998). Phosphogypsum efficiency to correct soil P deficiency and/or soil acidity. Unpublished report, 5p.

Slansky M 1986. Proterozoic and Cambrian phosphorites - regional overview: West Africa. In: Cook PJ and JH Shergold (eds.) Phosphate deposits of the world, Vol. 1: Proterozoic and Cambrian phosphorites, Cambridge University Press, Cambridge, UK:108-115.

Sustrac G 1986. BRGM phosphate prospecting methods and results in West Africa. *Trans. Instn. Min. Metall.* (Sect. A: Min. Industry) 95:A134-A143.

Witschard, F. (1965) -*Plan minéral* de la République du *Sénégal*. Dakar, Direction des Mines et de la Géologie, 725 pages.

Chapter 3

CLAY MINERALS

Clay minerals are hydrous aluminum phyllosilicates, sometimes with variable amounts of iron, magnesium, alkali metals, alkaline earths, and other cations. Clays have structures similar to the micas and therefore form flat hexagonal sheets. Clay minerals are common weathering products (including weathering of feldspar) and low temperature hydrothermal alteration products. Clay minerals are very common in fine grained sedimentary rocks such as shale, mudstone, and siltstone and in fine grained metamorphic slate and phyllite. Clays are ultrafine-grained (normally considered to be less than 2 micrometres in size on standard particle size classifications) and so require special analytical techniques. Standards include x-ray diffraction, electron diffraction methods, various spectroscopic methods such as Mossbauer spectroscopy, infrared spectroscopy, and SEM-EDS or automated mineralogy solutions. These methods can be augmenting polarized light microscopy, a traditional technique establishing fundamental occurrences or petrologic relationships. Clays are commonly referred to as 1:1 or 2:1. Clays are fundamentally built of tetrahedral sheets and octahedral sheets, as described in the structure section below.

A 1:1 clay would consist of one tetrahedral sheet and one octahedral sheet, and examples would be kaolinite and serpentine.

A 2:1 clay consists of an octahedral sheet sandwiched between two tetrahedral sheets, and examples are illite, smectite, attapulgite, and chlorite (although chlorite has an external octahedral sheet often referred to as "brucite").

Clay minerals include the following groups:

- Kaolin group which includes the minerals kaolinite, dickite, halloysite, and nacrite (polymorphs of $Al_2Si_2O_5(OH)_4$).
  - Some sources include the kaolinite-serpentine group due to structural similarities (Bailey, 1980).
- Smectite group which includes dioctahedral smectites such as montmorillonite and nontronite and trioctahedral smectites for example saponite.
- Illite group which includes the clay-micas. Illite is the only common mineral.
- Chlorite group includes a wide variety of similar minerals with considerable chemical variation.
- Other 2:1 clay types exist such as sepiolite or attapulgite, clays with long water channels internal to their structure.

Mixed layer clay variations exist for most of the above groups. Ordering is described as random or regular ordering, and is further described by the term reichweite, which is German for range or reach. Literature articles will refer to a R1 ordered illite-smectite, for example. This type would be ordered in an ISISIS fashion. R0 on the other hand describes random ordering, and other advanced ordering types are also found (R3, etc). Mixed layer clay minerals which are perfect R1 types often get their own names. R1 ordered chlorite-smectite is known as corrensite, R1 illite-smectite is rectorite.

REFERENCES

Bailey S.W (1980): Structures of layer silicates. In Crystal structures of clay mineral and theirX-ray identification. pp 1-124. Brindley and Brown Eds. *Mineralogical Society of London.*

Chapter 4

ZEOLITES AND GEOPOLYMER

SECTION A: ZEOLITES

What's Zeolite

Zeolites are well defined class of crystalline naturally occurring aluminosilicate minerals. They have three dimensional structures arising from a framework of $[SiO_4]^{4-}$ and $[AlO_4]^{4-}$ coordination polyhedra (Fig 1) linked by all their corners. The frameworks generally are very open and contain channels and cavities in which are located cations and water molecules (Fig 2). The cations often have a high degree of mobility giving rise to facile ion exchange and the water molecules are readily lost and regained: these accounts for the well-known dessicant properties of zeolites.

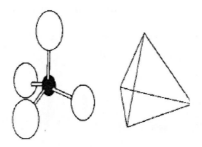

Figure 1. Representation of $[SiO_4]^{4-}$ or $[AlO_4]^{4-}$ tetrahedral.

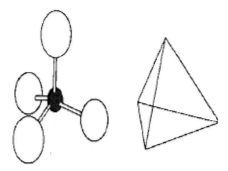

Figure 2. The framework structure of chabazites. Each line represents an oxygen atom and each junction a silicon or aluminium.

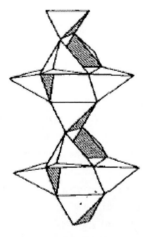

Figure 3. Tetrahedra linked together to create a three dimensional structure.

The word zeolite has Greek roots and means, an allusion to the visible loss of water noted when the natural zeolite are heated. This property of course, illustrates their easy water loss and is describe as 'intumescence'.

The Structure of Zeolites

Introduction

At present some 39 naturally occurring zeolites species have been recorded and their structures determined. In addition more than 100 synthetic species with no know natural counterparts have been reported in the literature.

Relatively few of these have been confirmed as new zeolites and the majority awaits full structural determination. This chapter will describe the major structural classes of know zeolites together with a more detailed consideration of those zeolites of current commercial significance.

Fundamental Zeolites Structural Units

As stated earlier all zeolites have framework (three-dimensional) structures constructed by joining together $[SiO_4]^{4-}$ and $[AlO_4]^{5-}$ coordination polyhedra. By definition these tetrahedra are assembled together such that the oxygen at each tetrahedral corner is shared with that in an identical tetrahedron (Si or Al), as shown in figure 3. This corner (or vertex) sharing creates infinite lattices comprised of identical building blocks (unit cells) in manner common to all crystalline materials.

Nature of Zeolite and Their Occurence

Zeolites of Saline Alkaline Lake Origin

The creation of zeolites in a saline alkaline lacustrine environment is typified by a closed basin in arid, or semi-arid, regions in which lake water of high sodium carbonate/bicarbonate content has gathered to produce a high pH (~9.5). Zeolites are laid down from reactive materials deposited in the lake. These materials are often "glasses" of volcanic origin, wind blown, and describes as "vitric tuffs". Other substances forming zeolites under these conditions are biogenic silica (skeletons of radiolareans and forams), clays, plagioclase (feldspar) and forms of quartz.

Deposits of this type commonly contain phillipsite, clinoptilolite and erionite, out some include chabazite and modernite. Zeolites locations of this sort are widespread in the Western USA stemming from the Plio-Pleistocene era. Older deposits (Eocene, Cenozoic) often contain analcime which arises from subsequent alteration of the zeolites in the younger rocks.

Zeolites in Soils and Land Surfaces

Locations of this environment are typified by the San Joaquim Valley, California, and the Big Sandy Formation in Arizona, USA, other well-studied sites are those in the Olduvai Gorge, Tanzania, and at lake Magadi in Kenya. Again dry, closed basins are required. In California the "reactive" material is mostly montmorillonite (a clay mineral) and the high pH is caused by evapotranspiration; analcime is the most abundant species. At the Olduvai

Gorge wind-blown tuffs have been altered to from phillipsite, natrolite, chabazite and analcime. Geologically these deposits are young (Pleistocene and Holocene), red-brown in colour and very abundant. The zeolites content of the soils in low (15-40%) and a similar occurrence is reported in the Harwell series of soil of Berkshire (UK), which have high heulandite content.

Zeolites in Marine Deposits

Marine zeolites readily form at shallow depths and low temperatures as well as at more substantial depths and higher temperatures. This latter condition will be discussed under burial diagenesis.

Zeolites from Open Flowing Systems

Zeolites can be formed when flowing waters of high pH and salt content interact with vitric volcanic ash causing rapid crystal formation. Sites demonstrating this genesis have been recorded in Nevada, in the Koko Crater (Hawaii) and in the abundant tuffs in southern Italy (typified by the yellow Neopolitan ashes). Tuffs in southern Italy commonly contain 60% chabazite with some 10% phillipsite.

Evidence suggests that time scales as short (by geological standard) as 4000 years are needed for these formations to occur. The high pH of the system stems from hydrolysis of the volcanic ash by surface water. Clinoptilolite, aanalcime, phillipsite and chabazite are found in these locations.

Hydrothermally Treated Zeolites

Zeolites are well known in Yellowstone Park (USA), in Iceland and in Wairakei (New Zealand) where they have formed from geothermal (geyser) action on existing volcanic ash.

Often deposits occur in well-defined zones, with clinoptilolite and mordenite forming in the shallowest and coolest zones with deeper (hotter) environments containing analcine, huelandite, laumontite and wairakite. Ferrierite, thomsonite, chabazita, misolite, solecite and stilbite are also known in hydrothermal zones.

Zeolites Formed by Burial Diagenesis

This classification relates to minerals formed as a result of their depth of burial, by subsequent layers of geologic species, and the consequential geothermal gradient. This group has strong associations with deep-sea ad hydrothermal conditions. If reflects decreases in hydration with increased depth, so relatively porous zeolites (clinoptilolite and mordenite) are found above those of a less open nature (analcime, heulandite and laumontite).

Deposits of this nature have been described in the Green Tuff region of Japan and in the John Day formation in Oregon, USA.

Summary

It can be appreciated that zeolites are readily formed in a variety of geological environments mainly from volcanic debris. They may well be important to the formation of other minerals (e.g. feldspars and clay minerals) by alteration and similar phenomena. Zeolites have increased in their geological standing during recent years-a story still progressing and enlivened by their association with some oil-bearing rocks and speculation as to the likelihood of zeolites structural cavities being suitable environments for the generation of protein precursors.

Zeolites Structure Identification and Characterization

X-Ray Methods

Ideally all zeolites structures should be capable of solution by the use of modern X-ray-single-crystal techniques and indeed naturally occurring species have been so studied. In the case of synthetic materials this requires the production of suitable single crystals. Despite special crystal growth techniques this is often not possible and has meant that a wide array of other techniques has been used for the investigation of zeolites structures.

Prominent amongst these is the sophisticated interpretation of X-ray powder diffraction data (XRD). In this method X-ray irradiation of zeolites powders (say 1-50μm crystallite diameter) produces a scattering pattern from regular arrays of atoms (or ions) within the structure. If reflects the framework and non-framework symmetry of the constituents of each zeolite to produce a diagnostic fingerprint of 2 θ (or'd') spacing according to the Bragg equation:

$$n\lambda = 2d \sin \theta$$

Where n is an integer, λ is the wavelength of the incident X-rays, d is the value of the interlayer spacings of the component atoms and ions and θ is the scattering angle.

These diagnostic patterns can provide an identification of known zeolite structures and Von Ballmoss has collected together a handbook (see bibliography) of computer-generated 'standard' patterns.

Table 1. Classification of zeolite structures [1]

Number of linked tetrahedra	SBU created	Shorthand description
4	4 oxygen ring	S4R
5	5 oxygen ring	S5R
6	6 oxygen ring	S6R
8	8 oxygen ring	S8R
8	4-4 oxygen ring	D4R
12	6-6 oxygen ring	D6R
16	8-8 oxygen ring	D8R

[1] (S = single ring, D = double, R = ring)

Zeolite Syntheses

Introduction

Generally the starting point for zeolite synthesis is crystallization from an inhomogeneous gel, created from a silica source and an alumina source, combined with water under high pH conditions generated by OH⁻ ion concentrations. Control of SiO_2: $Al_2 O_3$ ratio in this gel qualifies the final framework composition of the product and usually all the aluminium available is taken into the zeolite final composition (a notable exception to this is zeolite A).

Components for Zeolite Synthesis

Sources of Aluminium

Zeolite production from the aluminium source is enhanced by the presence of the $[Al(OH)_4]$ moiety. The optimal pH range for successful is mainly pH 11-13, although some variation with zeolite species occurs outside this range.

Sources of Silicon

The most widely used sources are soluble silicates and their hydrates (e.g. sodium metasilicate pentahydrate), silica sols (e.g. 30% by weight SiO_2) made from high surface area ('funded') silicas such as Cab-o-Sil and commercial products (e.g. Syton, Ludox, etc.) less frequently used are silica gels and glasses (including volcanic glass) silicon esters, clays (e.g. Kaolinite), volcanic tuffs, sand and quartz.

Sources of Cations

Obviously if high pH favours zeolite crystallization then alkali metals and alkaline earth hydroxides are the preferred cation source in most cases-certainly in industrial zeolite production. Other oxides and some salts have been used and, of course, the soluble silicates and aluminates quoted above are themselves cation sources.

Often mixtures of cations have been used and early work by Barrer and Denny (1961) used an organic cation-tetrathylammonium (TMA). Since their work many syntheses including organic cations have been developed and they are particularly useful in the syntheses of high-silica zeolites (e.g. ZSM-5), phases of silica with zeolitic properties (e.g. silicalite) and some zeotypes (e.g. the AIPO$_4$ materials).

Other Materials in Zeolite Synthesis Mixtures

Quite apart from organic bases many different organic compounds have profitably been included in syntheses mixtures. A whole range of zeolites have been prepared from gels containing organic dyestuffs. The role that these materials and cations play in crystallization will be discussed in the last section of this chapter.

Reaction Variables in Zeolite Synthesis

Concentration, Temperature, Pressure and Time

Usually reaction mixtures are composed of the appropriate sources to give the required Si: Al zeolite framework composition with an excess of hydroxide present. As to the OH$^-$: SiO$_2$ ratio is increased more silicate remains in solution and lower Si: Al products are formed. Obviously the H$_2$O:SiO$_2$ ratio is inherently linked to this as well.

Zeolites of lower water content have traditionally been prepared at higher temperatures (up to 350°C) and pressures (developed autogeneously in sealed reaction vessels). Actually these denser zeolite phases (e.g. analcime) can be favoured at high pH and increasingly this knowledge is being used to prepare even the dense zeolites under relatively mild hydrothermal conditions. Recent literature quotes low-temperature hydrothermal preparations of mordenite, ZSM-5 and analcime; these tend to be recipes using organic bases.

Table 1. Typical oxide formula of some synthetic zeolite

Zeolites	Typical oxide formula
Zeolites A	$Na_2.Al_2O_3.2SiO_2.4,5H_2O$
Zeolites N-A	$(Na,TMA)_2O.Al_2O_3.4,8SiO_2 7H_2O$ TMA-$(CH3)_4 N^+$
Zeolites H	$K_2O.Al_2O_3.2SiO_2.4H_2O$
Zeolites L	$(K_2Na_2)O. Al_2O_3 6SiO_2 5H_2O$
Zeolites X	$Na_2O.Al_2O_3 2,5SiO_2 6H_2O$
Zeolites Y	$Na_2O. Al_2O_3.4,8 SiO_2.8,9H_2O$
Zeolites P	$Na_2O. Al_2O_3 2 - 5SiO_2 5H_2O$
Zeolites O	$(Na_2,K_2,TMA_2)O.Al_2O_3 7SiO_2.3,5 H_2O;TMA - (CH_3)_4N^+$
Zeolites Ω	$(Na,TMA)_2O.Al_2O_3 7SiO_2.5H_2O;TMA - (CH_3)_4N^+$
Zeolites ZK-4	$0,85Na_2O.0,15 (TMA)_2O. Al_2O_3 3,3SiO_2.6H_2O$
Zeolites ZK-5	$(R,Na_2)O.Al_2O_3.4-6SiO_2.6H_2O$

The element of time is important in two ways:

(i) an induction period during which the reaction mixture is held near ambient temperature prior to raising to the crystallization temperature often optimizes zeolite yield (as in X, Y syntheses);

(ii) often different zeolites crystallize from one reaction mixture at different times. This second time element is because all zeolites are metastable species and in nature many examples are know of diagenetic sequences whereby more open zeolite structures (e.g. phillipsite) convert, over a geological time span, to less open structures (e.g. clinoptilolite) and finally to analcime—the most stable and dense of common zeolites. On a laboratory, or plant, scale crystallization times are important to the production of A, X and Y which are metastable to NaP, so industrially reaction mixtures are quenched at optimal crystallization times. Similarly modernite can transform to analcime.

Other Parameters

Many other factors can influence zeolite synthesis. Several studies record the effects of small amounts of salt impurities (iron salts need to be excluded from synthesis mixtures) and the consequences of trace amounts of aluminium in silicia sources. Others concentrate on the mode and nature of stirring and the influence of the order of addition of reactants together with the nature of the reaction vessels. Each factor can have an effect but no broad generalization can be drawn to help in the understanding of the mecanisms of crystallization involved, except to note that glass reaction vessels demonstrate memory effects and can drastically alter the course of syntheses. Presumably this comes from tiny seed crystals held in the glass surfaces etched by the high pH reaction mixtures. Plastic vessels or liners are advisable and even these must be very thoroughly cleaned before reuse.

Kinetics and Mechanisms in Zeolite Crystallizations

Precursors

Currently, there is no comprehensive explanation of routes whereby three-dimensional aluminosilicate structures grow from reaction mixtures to create zeolite frameworks. The most obvious missing step is an accurate knowledge of how different secondary bulding units (sbus) arise from precursor species. Clearly the $[Al(OH)_4]$ species plays an important part, being a tetrahedral entity known to be present at high solution pH, but its reaction with silicate species to form aluminium in the silicate sources used to make synthetic faujasites (i.e. X, Y) can be critical to their ease of production. Modern NMR and Raman spectroscopy have identified ring silicate and aluminosilicate anions in solution although information on zeolite-forming gels is sparse. This can be expected to be expanded as modern solid-state NMR and IR studies progress.

Given that precursors exist, the next problem encountered is to explain the agencies whereby the precursors grow into differing structural frameworks, i.e. what are the structure-directing entities present in zeolite reaction mixtures? A longstanding concept is that this arises from a templating action controlled by the cations present. This correspond, at least simplistically, to the accepted concept that there are two types of ion in solution —one which disrupts the existing water structure (structure-breaking) and one whose presence increases the existing water structure (structure forming). Na^+ is an example of a structure-breaking ion and Ca^{2+} is a structure-forming ion. Another casual observation is that some sbus seem to be stabilized in synthesis by specific cations. Sodium ions have been said to promote structures containing D4R and

D6R units, sodalite and gmelinite cages, whereas K, Ba an Rb promote cancrinite cages. However no similar comment can be directed to zeolite structures containing single ring sbus.

Conclusion

In the interpretation of the consolidation mechanism we refer explicitly to the theory of zeolitic material synthesis rather than to geopolymers. The major difference between geopolymers and zeolite is that zeolite material are mainly composed of crystal in closed cage structures, while the geopolymerization consist in the formation of a 3D aluminosilicate amorphous gel/polymer, closer to what authors observe here. However, for geopolymer, the phases observed are badly crystallized and resemble more to a gel. What is however new is that currently people who make geopolymer generally use metakaolin which is strongly reactive with soda and sodium silicates.

SECTION B: GEOPOLYMER

Geopolymer is a term covering a class of synthetic aluminosilicate materials with potential use in a number of areas, but predominantly as a replacement for Portland cement. The name Geopolymer was first applied to these materials by Joseph Davidovits in the 1970s, although similar materials had been developed in the former Soviet Union since the 1950s under the name Soil cements.

Research

Much of the drive behind research is to investigate the development of geopolymers as a potential large-scale replacement for concrete produced from Portland cement. This is due to geopolymers' lower carbon dioxide emissions, greater chemical and thermal resistance and better mechanical properties at both atmospheric and extreme conditions.

Production

Geopolymers are generally formed by reaction of an aluminosilicate powder with an alkaline silicate solution at roughly ambient conditions. Metakaolin is a commonly used starting material for laboratory synthesis of geopolymers, and is generated by thermal activation of kaolinite clay. Geopolymers can also be made from natural sources of pozzolanic materials, such as lava or fly ash from coal. Most studies have been carried out using natural or industrial waste sources of metakaolin and other aluminosilicates.

Theory

The majority of the Earth's crust is made up of Si-Al compounds. Davidovits proposed in 1978 that a single aluminium and silicon-containing compound, most likely geological in origin, could react in a polymerisation process with an alkaline solution. The binders created were termed "geopolymers" but, now, the majority of aluminosilicate sources are by-products from organic combustion, such as fly ash from coal burning. These inorganic polymers have a chemical composition somewhat similar to zeolitic materials but exist as amorphous solids, rather than having a crystalline microstructure.

Structure

The chemical reaction that takes place to form geopolymers follows a multi-step process:

1. Dissolution of Si and Al atoms from the source material due to hydroxide ions in solution,
2. Reorientation of precursor ions in solution, and
3. Setting via polycondensation reactions into an inorganic polymer.

The inorganic polymer network is in general a highly-coordinated 3-dimensional aluminosilicate gel, with the negative charges on tetrahedral Al(III) sites charge-balanced by alkali metal cations.

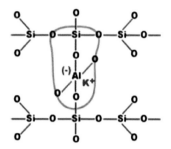

Figure 4. Structure of a geopolymer.

History

Davidovits has combined his expertise in alumino-silicate chemistry with a long-standing interest in archeology, particularly the archeology of ancient Egypt, and his examination of the building blocks of some of the major pyramids have led him to the conclusion that, rather than being blocks of solid limestone hauled into position, they are composed of geopolymers, cast in their final positions in the structure. He also considers that roman cement and the small artifacts, previously thought to be stone, of the Tiahuanaco civilisation were made using knowledge of geopolymer techniques.

References

Barrer R. M. and Denny P. J.; 1961 "Hydrothermal chemistry of the silicates". Part X. A partial study of the field CaO–Al2O3–SiO2–H2O J. Chem. Soc., , 983-1000, DOI: 10.1039/JR9610000983.

Davidovits, Joseph; Morris, Margie (1988). The pyramids: an enigma solved. New York: Hippocrene Books. ISBN 0 87052 559 X.

Davidovits, Joseph; Aliaga, Francisco (1981). Fabrication of stone objects, by geopolymeric synthesis, in the pre-incan Huanka civilization (Peru). *Making Cements with Plant Extracts.* Geopolymer Institute. Retrieved on 2008-01-09.

Dimitar Geogiev, Bogdan Bogdanov, Krasimira Angelova, Irena Markovska, Yancho Hristov (2009)" Synthetic zeolites-Structure, Classification, current trends in zeolite synthesis review, International science conference, 4-5 jui,, Stara Zagora, Bulgaria" Economics and Society development on the Base of Knowledge"

Dyer A. (1988), An introduction to zeolite molecular sieves. New York: John Wiley and Sons.

Roland von Ballmoos, 1981, The O-18-exchange method in zeolite chemistry synthesis, characterization and dealumination of high silica zeolites 10.3929/ethz-a-000215092.

Stabilization/solidification of hazardous and radioactive wastes with alkali-activated cements Science Direct *Journal of Hazardous Materials* 2005-08-13.

Geopolymer technology: the current state of the art *Journal of Materials Science*, 2006-06-04.

Chapter 5

PORTLAND CEMENT
AND POUZZOLANIC REACTION

SECTION A: FABRICATION OF PORTLAND CEMENT

Portland cement is the most common type of cement in general usage in many parts of the world, as it is a basic ingredient of concrete, mortar, stucco and most non-specialty grout. It is a fine powder produced by grinding Portland cement clinker (more than 90%), a limited amount of calcium sulfate which controls the set time, and up to 5% minor constituents (as allowed by various standards). As defined by the European Standard EN197.1, "Portland cement clinker is a hydraulic material which shall consist of at least two-thirds by mass of calcium silicates ($3CaO.SiO_2$ and $2CaO.SiO_2$), the remainder consisting of aluminium and iron-containing clinker phases and other compounds. The ratio of CaO to SiO_2 shall not be less than 2.0. The magnesium content (MgO) shall not exceed 5.0% by mass." (The last two requirements were already set out in the German Standard, issued in 1909)".

Portland cement clinker is made by heating, in a kiln, a homogeneous mixture of raw materials to a sintering temperature, which is about 1450°C for modern cements. The aluminium oxide and iron oxide are present as a flux and contribute little to the strength. For special cements, such as Low Heat (LH) and Sulfate Resistant (SR) types, it is necessary to limit the amount of tricalcium aluminate ($3CaO.Al_2O_3$) formed. The major raw material for the clinker-making is usually limestone ($CaCO_3$) mixed with a second materials containing clay as source of alumino-silicate. Normally, an impure limestone which contains clay or SiO_2 is used. The $CaCO_3$ content of these limestones can be as low as 80%. Second raw materials (materials in the rawmix other than limestone) depend on the purity of the limestone. Some of the second raw materials used are: clay, shale, sand, iron ore, bauxite, fly ash and slag. When a cement kiln is fired by coal, the ashes of the coal act as a secondary raw material.

History

Portland was developed from cements (or correctly hydraulic limes) made in Britain in the early part of the nineteenth century, and its name is derived from its similarity to Portland stone, a type of building stone that was quarried on the Isle of Portland in Dorset, England. Joseph Aspdin, a British bricklayer, in 1824 was granted a patent for a process of making a cement which he called Portland cement. His cement was an artificial hydraulic lime similar in properties to the material known as "Roman Cement" (patented in 1796 by James Parker) and his process was similar to that patented in 1822 and used since 1811 by James Frost who called his cement "British Cement". The name "Portland cement" is also recorded in a directory published in 1823 being associated with a William Lockwood and possibly others. Aspdin's son William in 1843 made an improved version of this cement and he initially called it "Patent Portland cement" although he had no patent. In 1848 William Aspdin further improved his cement and in 1853 moved to Germany where he was involved in cement making. Many people have claimed to have made the first Portland cement in the modern sense, but it is generally accepted that it was first manufactured by William Aspdin at Northfleet, England in about 1842. The German Government issued a standard on Portland cement in 1878.

Production

There are three fundamental stages in the production of Portland cement:

1. Preparation of the raw mixture
2. Production of the clinker
3. Preparation of the cement

The chemistry of cement is very complex, so cement chemist notation was invented to simplify the formula of common oxides found in cement. This reflects the fact that most of the elements are present in their highest oxidation state, and chemical analyses of cement are expressed as mass percent of these notional oxides.

Rawmix Preparation

The raw materials for Portland cement production are a mixture (as fine powder in the 'Dry process' or in the form of a slurry in the 'Wet process') of minerals containing calcium oxide, silicon oxide, aluminium oxide, ferric oxide, and magnesium oxide. The raw materials are usually quarried from local rock, which in some places is already practically the desired composition and in other places requires the addition of clay and limestone, as well as iron ore, bauxite or recycled materials. The individual raw materials are first crushed, typically to below 50 mm. In many plants, some or all of the raw materials are then roughly blended in a "prehomogenization pile". The raw materials are next ground together in a rawmill. Silos of individual raw materials are arranged over the feed conveyor belt. Accurately controlled proportions of each material are delivered onto the belt by weigh-feeders. Passing into the rawmill, the mixture is ground to rawmix. The fineness of rawmix is specified in terms of the size of the largest particles, and is usually controlled so that there are less than 5-15% by mass of particles exceeding 90 μm in diameter. It is important that the rawmix contains no large particles in order to complete the chemical reactions in the kiln, and to ensure the mix is chemically homogenous. In the case of a dry process, the rawmill also dries the raw materials, usually by passing hot exhaust gases from the kiln through the mill, so that the rawmix emerges as a fine powder. This is conveyed to the blending system by conveyor belt or by a powder pump. In the case of wet process, water is added to the rawmill feed, and the mill product is a slurry with moisture content usually in the range 25-45% by mass. This slurry is conveyed to the blending system by conventional liquid pumps.

Rawmix Blending

The rawmix is formulated to a very tight chemical specification. Typically, the content of individual components in the rawmix must be controlled within 0.1% or better. Calcium and silicon are present in order to form the strength-producing calcium silicates. Aluminium and iron are used in order to produce liquid ("flux") in the kiln burning zone. The liquid acts as a solvent for the silicate-forming reactions, and allows these to occur at an economically low temperature. Insufficient aluminium and iron lead to difficult burning of the clinker, while excessive amounts lead to low strength due to dilution of the silicates by aluminates and ferrites. Very small changes in calcium content lead to large changes in the ratio of alite to belite in the clinker, and to corresponding changes in the cement's strength-growth characteristics. The relative amounts of each oxide are therefore kept constant in order to maintain steady conditions in the kiln, and to maintain constant product properties. In practice, the rawmix is controlled by frequent chemical analysis (hourly by X-Ray fluorescence analysis, or every 3 minutes by prompt gamma neutron activation analysis). The analysis data is used to make automatic adjustments to raw material feed rates. Remaining chemical variation is minimized by passing the raw mix through a blending system that homogenizes up to a day's supply of rawmix (15,000 tonnes in the case of a large kiln).

Formation of Clinker

Typical Clinker Nodules

The raw mixture is heated in a cement kiln, a slowly rotating and sloped cylinder, with temperatures increasing over the length of the cylinder up to a peak temperature of 1400-1450°C. A complex succession of chemical reactions take place (see cement kiln) as the temperature rises. The peak temperature is regulated so that the product contains sintered but not fused lumps. Sintering consists of the melting of 25-30% of the mass of the material. The resulting liquid draws the remaining solid particles together by surface tension, and acts as a solvent for the final chemical reaction in which alite is formed. Too low a temperature causes insufficient sintering and incomplete reaction, but too high a temperature results in a molten mass or glass, destruction of the kiln lining, and waste of fuel. When all goes to plan, the resulting material is clinker. On cooling, it is conveyed to storage. Some effort is usually made to blend the clinker, because although the chemistry of the rawmix may have been tightly controlled, the kiln process potentially introduces new sources of chemical variability. The clinker can be stored for a number of years before use. Prolonged exposure to water decreases the reactivity of cement produced from weathered clinker.

The enthalpy of formation of clinker from calcium carbonate and clay minerals is ~1700 kJ/kg. However, because of heat loss during production, actual values can be much higher. The high energy requirements and the release of significant amounts of carbon dioxide make cement production a concern for global warming. See "Environmental effects" below.

Cement Grinding

In order to achieve the desired setting qualities in the finished product, a quantity (2-8%, but typically 5%) of calcium sulfate (usually gypsum or anhydrite) is added to the clinker and the mixture is finely ground to form the finished cement powder. This is achieved in a cement mill. The grinding process is controlled to obtain a powder with a broad particle size range, in which typically 15% by mass consists of particles below 5 μm diameter, and 5% of particles above 45 μm.

Table 1. Typical constituents of Portland clinker and Portland cement. Cement industry style notation in italics:

Clinker	Mass%	Cement	Mass%
Tricalcium silicate $(CaO)_3.SiO_2$, *C₃S*	45-75%	Calcium oxide, CaO, *C*	61-67%
Dicalcium silicate $(CaO)_2.SiO_2$, *C₂S*	7-32%	Silicon oxide, SiO_2, *S*	19-23%
Tricalcium aluminate $(CaO)_3.Al_2O_3$, *C₃A*	0-13%	Aluminium oxide, Al_2O_3, *A*	2.5-6%
Tetracalcium aluminoferrite $(CaO)_4.Al_2O_3.Fe_2O_3$, *C₄AF*	0-18%	Ferric oxide, Fe_2O_3, *F*	0-6%
Gypsum $CaSO_4$ 2 H_2O	2-10%	Sulfate	

The measure of fineness usually used is the "specific surface", which is the total particle surface area of a unit mass of cement. The rate of initial reaction (up to 24 hours) of the cement on addition of water is directly proportional to the specific surface. Typical values are 320-380 $m^2 \cdot kg^{-1}$ for general purpose cements, and 450-650 $m^2 \cdot kg^{-1}$ for "rapid hardening" cements. The cement is conveyed by belt or powder pump to a silo for storage. Cement plants normally have sufficient silo space for 1-20 weeks production, depending upon local demand cycles. The cement is delivered to end-users either in bags or as bulk powder blown from a pressure vehicle into the customer's silo. In developed countries, 80% or more of cement is delivered in bulk, and many cement plants have no bag-packing facility. In developing countries, bags are the normal mode of delivery.

Use

The most common use for Portland cement is in the production of concrete. Concrete is a composite material consisting of aggregate (gravel and sand), cement, and water. As a construction material, concrete can be cast in almost any shape desired, and once hardened, can become a structural (load bearing) element. Users may be involved in the factory production of pre-cast units, such as panels, beams, road furniture, or may make cast-*in-situ* concrete such as building superstructures, roads, dams. These may be supplied with concrete mixed on site, or may be provided with "ready-mixed" concrete made at permanent mixing sites. Portland cement is also used in mortars (with sand and water only) for plasters and screeds, and in grouts (cement/water mixes squeezed into gaps to consolidate foundations, road-beds, etc).

Setting and Hardening

When water is mixed with Portland cement, the product sets in a few hours and hardens over a period of weeks. These processes can vary widely depending upon the mix used and the conditions of curing of the product, but a typical concrete sets (i.e. becomes rigid) in about 6 hours, and develops a compressive strength of 8~ MPa in 24 hours. The strength rises to 15~ MPa at 3 days, 23~ MPa at one week, 35~ MPa at 4 weeks, and 41~ MPa at three months. In principle, the strength continues to rise slowly as long as water is available for continued hydration, but concrete is usually allowed to dry out after a few weeks, and this causes strength growth to stop.

Setting and hardening of Portland cement is caused by the formation of water-containing compounds, forming as a result of reactions between cement components and water. Usually, cement reacts in a plastic mixture only at water/cement ratios between 0.25 and 0.75. The reaction and the reaction products are referred to as hydration and hydrates or hydrate phases, respectively. As a result of the reactions (which start immediately), a stiffening can be observed which is very small in the beginning, but which increases with time. The point in time at which it reaches a certain level is called the start of setting. The consecutive further consolidation is called setting, after which the phase of hardening begins.

Stiffening, setting and hardening are caused by the formation of a microstructure of hydration products of varying rigidity which fills the water-filled interstitial spaces between the solid particles of the cement paste, mortar or concrete. The behaviour with time of the stiffening, setting and hardening therefore depends to a very great extent on the size of the interstitial spaces, i.

e. on the water/cement ratio. The hydration products primarily affecting the strength are calcium silicate hydrates ("C-S-H phases"). Further hydration products are calcium hydroxide, sulfatic hydrates (AFm and AFt phases), and related compounds, hydrogarnet, and gehlenite hydrate. Calcium silicates or silicate constituents make up over 70 % by mass of silicate-based cements. The hydration of these compounds and the properties of the calcium silicate hydrates produced are therefore particularly important. Calcium silicate hydrates contain less CaO than the calcium silicates in cement clinker, so calcium hydroxide is formed during the hydration of Portland cement. This is available for reaction with supplementary cementitious materials such as ground granulated blast furnace slag and pozzolans. The simplified reaction of alite with water may be expressed as:

$$2Ca_3OSiO_4 + 6H_2O \rightarrow 3CaO.2SiO_2.3H_2O + 3Ca(OH)_2$$

This is a relatively fast reaction, causing setting and strength development in the first few weeks. The reaction of belite is:

$$2Ca_2SiO_4 + 4H_2O \rightarrow 3CaO.2SiO_2.3H_2O + Ca(OH)_2$$

This reaction is relatively slow, and is mainly responsible for strength growth after one week. Tricalcium aluminate hydration is controlled by the added calcium sulfate, which immediately goes into solution when water is added. Firstly, ettringite is rapidly formed, causing a slowing of the hydration (see tricalcium aluminate):

$$Ca_3(AlO_3)_2 + 3CaSO_4 + 32H_2O \rightarrow Ca_6(AlO_3)_2(SO_4)_3.32H_2O$$

The ettringite subsequently reacts slowly with further tricalcium aluminate to form "monosulfate" - an "AFm phase":

$$Ca_6(AlO_3)_2(SO_4)_3.32H_2O \quad + \quad Ca_3(AlO_3)_2 \quad + \quad 4H_2O \quad \rightarrow$$
$$3Ca_4(AlO_3)_2(SO_4).12H_2O$$

This reaction is complete after 1-2 days. The calcium aluminoferrite reacts slowly due to precipitation of hydrated iron oxide:

$$2Ca_2AlFeO_5 + CaSO_4 + 16H_2O \rightarrow Ca_4(AlO_3)_2(SO_4).12H_2O + Ca(OH)_2 + 2Fe(OH)_3$$

The pH-value of the pore solution reaches comparably high values and is of importance for most of the hydration reactions.

Soon after Portland cement is mixed with water, a brief and intense hydration starts (pre-induction period). Calcium sulfates dissolve completely and alkali sulfates almost completely. Short, hexagonal needle-like ettringite crystals form at the surface of the clinker particles as a result of the reactions between calcium- and sulfate ions with tricalcium aluminate. Further, originating from tricalcium silicate, first calcium silicate hydrates (C-S-H) in colloidal shape can be observed. Caused by the formation of a thin layer of hydration products on the clinker surface, this first hydration period ceases and the induction period starts during which almost no reaction takes place. The first hydration products are too small to bridge the gap between the clinker particles and do not form a consolidated microstructure. Consequently the mobility of the cement particles in relation to one another is only slightly affected; i. e. the consistency of the cement paste turns only slightly thicker. Setting starts after approximately one to three hours, when first calcium silicate hydrates form on the surface of the clinker particles, which are very fine-grained in the beginning. After completion of the induction period, a further intense hydration of clinker phases takes place. This third period (accelerated period) starts after approximately four hours and ends after 12 to 24 hours. During this period a basic microstructure forms, consisting of C-S-H needles and C-S-H leafs, platy calcium hydroxide and ettringite crystals growing in longitudinal shape. Due to growing crystals, the gap between the cement particles is increasingly bridged. During further hydration, the hardening steadily increases, but with decreasing speed. The density of the microstructure rises and the pores fill: the filling of pores causes strength gain.

Types of Portland Cement

General

There are different standards for classification of Portland cement. The two major standards are the ASTM C150 used primarily in the U.S. and European EN-197. EN 197 cement types CEM I, II, III, IV, and V do not correspond to the similarly-named cement types in ASTM C 150.

ASTM C150

There are five types of Portland cements with variations of the first three according to ASTM C150.

Type I Portland cement is known as common or general purpose cement. It is generally assumed unless another type is specified. It is commonly used for general construction especially when making precast and precast-prestressed concrete that is not to be in contact with soils or ground water. The typical compound compositions of this type are:

55% (C_3S), 19% (C_2S), 10% (C_3A), 7% (C_4AF), 2.8% MgO, 2.9% (SO_3), 1.0% Ignition loss, and 1.0% free CaO.

A limitation on the composition is that the (C_3A) shall not exceed fifteen percent.

Type II is intended to have moderate sulfate resistance with or without moderate heat of hydration. This type of cement costs about the same as Type I. Its typical compound composition is:

51% (C_3S), 24% (C_2S), 6% (C_3A), 11% (C_4AF), 2.9% MgO, 2.5% (SO_3), 0.8% Ignition loss, and 1.0% free CaO.

A limitation on the composition is that the (C_3A) shall not exceed eight percent which reduces its vulnerability to sulfates. This type is for general construction that is exposed to moderate sulfate attack and is meant for use when concrete is in contact with soils and ground water especially in the western United States due to the high sulfur content of the soil. Because of similar price to that of Type I, Type II is much used as a general purpose cement, and the majority of Portland cement sold in North America meets this specification.

Note: Cement meeting (among others) the specifications for Type I and II has become commonly available on the world market.

Type III has relatively high early strength. Its typical compound composition is:

57% (C_3S), 19% (C_2S), 10% (C_3A), 7% (C_4AF), 3.0% MgO, 3.1% (SO_3), 0.9% Ignition loss, and 1.3% free CaO.

This cement is similar to Type I, but ground finer. Some manufacturers make a separate clinker with higher C_3S and/or C_3A content, but this is increasingly rare, and the general purpose clinker is usually used, ground to a specific surface typically 50-80% higher. The gypsum level may also be increased a small amount. This gives the concrete using this type of cement a three day compressive strength equal to the seven day compressive strength of types I and II. Its seven day compressive strength is almost equal to types I and II 28 day compressive strengths. The only downside is that the six month strength of type III is the same or slightly less than that of types I and II. Therefore the long-term strength is sacrificed a little. It is usually used for precast concrete manufacture, where high 1-day strength allows fast turnover of molds. It may also be used in emergency construction and repairs and construction of machine bases and gate installations.

Type IV Portland cement is generally known for its low heat of hydration. Its typical compound composition is:

28% (C_3S), 49% (C_2S), 4% (C_3A), 12% (C_4AF), 1.8% MgO, 1.9% (SO_3), 0.9% Ignition loss, and 0.8% free CaO.

The percentages of (C_2S) and (C_4AF) are relatively high and (C_3S) and (C_3A) are relatively low. A limitation on this type is that the maximum percentage of (C_3A) is seven, and the maximum percentage of (C_3S) is thirty-five. This causes the heat given off by the hydration reaction to develop at a slower rate. However, as a consequence the strength of the concrete develops slowly. After one or two years the strength is higher than the other types after full curing. This cement is used for very large concrete structures, such as dams, which have a low surface to volume ratio. This type of cement is generally not stocked by manufacturers but some might consider a large special order. This type of cement has not been made for many years, because Portland-pozzolan cements and ground granulated blast furnace slag addition offer a cheaper and more reliable alternative.

Type V is used where sulfate resistance is important. Its typical compound composition is:

38% (C_3S), 43% (C_2S), 4% (C_3A), 9% (C_4AF), 1.9% MgO, 1.8% (SO_3), 0.9% Ignition loss, and 0.8% free CaO.

This cement has a very low (C_3A) composition which accounts for its high sulfate resistance. The maximum content of (C_3A) allowed is five percent for Type V Portland cement. Another limitation is that the (C_4AF) + 2(C_3A) composition cannot exceed twenty percent. This type is used in concrete that is to be exposed to alkali soil and ground water sulfates which react with (C_3A) causing disruptive expansion.

It is unavailable in many places although its use is common in the western United States and Canada. As with Type IV, Type V Portland cement has mainly been supplanted by the use of ordinary cement with added ground granulated blast furnace slag or tertiary blended cements containing slag and fly ash. Types Ia, IIa, and IIIa have the same composition as types I, II, and III. The only difference is that in Ia, IIa, and IIIa an air-entraining agent is ground into the mix.

I	Portland cement	Comprising Portland cement and up to 5% of minor additional constituents
II	Portland-composite cement	Portland cement and up to 35% of other single constituents
III	Blastfurnace cement	Portland cement and higher percentages of blastfurnace slag
IV	Pozzolanic cement	Portland cement and up to 55% of pozzolanic constituents
V	Composite cement	Portland cement, blastfurnace slag and pozzolana or fly ash

The air-entrainment must meet the minimum and maximum optional specification found in the ASTM manual. These types are only available in the eastern United States and Canada but can only be found on a limited basis. They are a poor approach to air-entrainment which improves resistance to freezing under low temperatures.

EN 197

EN 197-1 defines 5 classes of common cement that comprise Portland cement as a main constituent. These classes differ from the ASTM classes.

Constituents that are permitted in Portland-composite cements are blastfurnace slag, silica fume, natural and industrial pozzolans, silicious and calcareous fly ash, burnt shale and limestone.

White Portland Cement

White Portland cement differs physically from the gray form only in its color, and as such can fall into many of the above categories (e.g. ASTM Type I, II and/or III). However, its manufacture is significantly different from that of the gray product, and is treated separately.

Safety and Environmental Effects

Safety

When cement is mixed with water a highly alkaline solution (pH ~13) is produced by the dissolution of calcium, sodium and potassium hydroxides. Gloves, goggles and a filter mask should be used for protection. Hands should be washed after contact. Cement can cause serious burns if contact is prolonged or if skin is not washed promptly. Once the cement hydrates, the hardened mass can be safely touched without gloves.

In Scandinavia, France and the UK, the level of chromium (VI), which is thought to be toxic and a major skin irritant, may not exceed 2 ppm (parts per million).

Environmental Effects

Portland cement manufacture can cause environmental impacts at all stages of the process. These include emissions of airborne pollution in the form of dust, gases, noise and vibration when operating machinery and during blasting in quarries, consumption of large quantities of fuel during manufacture, release of CO_2 from the raw materials during manufacture, and damage to countryside from quarrying. Equipment to reduce dust emissions during quarrying and manufacture of cement is widely used, and equipment to trap and separate exhaust gases are coming into increased use. Environmental protection also includes the re-integration of quarries into the countryside after they have been closed down by returning them to nature or re-cultivating them.

Epidemiologic Notes and Reports Sulfur Dioxide Exposure in Portland Cement Plants, from the Centers for Disease Control states "Workers at Portland cement facilities, particularly those burning fuel containing sulfur, should be aware of the acute and chronic effects of exposure to SO_2 [sulfur dioxide], and peak and full-shift concentrations of SO_2 should be periodically measured. The Arizona Department of Environmental Quality was informed this week that the Arizona Portland Cement Co. failed a second round of testing for emissions of hazardous air pollutants at the company's Rillito plant near Tucson. The latest round of testing, performed in January 2003 by the company, is designed to ensure that the facility complies with federal standards governing the emissions of dioxins and furans, which are byproducts of the manufacturing process. Cement Reviews' "Environmental News" web page details case after case of environmental problems with cement manufacturing.

An independent research effort of AEA Technology to identify critical issues for the cement industry today concluded the most important environment, health and safety performance issues facing the cement industry are atmospheric releases (including greenhouse gas emissions, dioxin, NO_x, SO_2, and particulates), accidents and worker exposure to dust.

The CO_2 associated with Portland cement manufacture falls into 3 categories:

(1) CO_2 derived from decarbonation of limestone,
(2) CO_2 from kiln fuel combustion,
(3) CO_2 produced by vehicles in cement plants and distribution.

Source 1 is fairly constant: minimum around 0.47 kg CO_2 per kg of cement, maximum 0.54, typical value around 0.50 world-wide. Source 2 varies with plant efficiency: efficient precalciner plant 0.24 kg CO_2 per kg cement, low-efficiency wet process as high as 0.65, typical modern practices (e.g UK) averaging around 0.30. Source 3 is almost insignificant at 0.002-0.005. So typical total CO_2 is around 0.80 kg CO_2 per kg finished cement. This leaves aside the CO_2 associated with electric power consumption, since this varies according to the local generation type and efficiency. Typical electrical energy consumption is of the order of 90-150 kWh per tonne cement, equivalent to 0.09-0.15 kg CO_2 per kg finished cement if the electricity is coal-generated.

Overall, with nuclear- or hydroelectric power and efficient manufacturing, CO_2 generation can be as little as 0.7 kg per kg cement, but can be as high as twice this amount. The thrust of innovation for the future is to reduce sources 1 and 2 by modification of the chemistry of cement, by the use of wastes, and by adopting more efficient processes. Although cement manufacturing is clearly a very large CO_2 emitter, concrete (of which cement makes up about 15%) compares quite favourably with other building systems in this regard. See also cement kiln emissions.

Cement Plants as Alternatives to Conventional Waste Disposal or Processing

Due to the high temperatures inside cement kilns, combined with the oxidizing (oxygen-rich) atmosphere and long residence times, cement kilns have been used as a processing option for various types of waste streams. The waste streams often contain combustible material which allows the substitution of part of the fossil fuel normally used in the process.

Waste materials used in cement kilns as a fuel supplement:

1. Car and truck tires; steel belts are easily tolerated in the kilns
2. Waste solvents and lubricants.
3. Hazardous waste; cement kilns completely destroy hazardous organic compounds
4. Bone meal; slaughter house waste due to bovine spongiform encephalopathy contamination concerns
5. Waste plastics
6. Sewage sludge
7. Rice shells
8. Sugar cane waste

Portland cement manufacture also has the potential to remove industrial by-products from the waste-stream, effectively sequestering some environmentally damaging wastes. These include:

1. Slag
2. Fly ash (from power plants)
3. Silica fume (from steel mills)
4. Synthetic gypsum (from desulfurisation)

SECTION B: MANUFACTURE AND USES OF LIME POZZOLANIC REACTION

Manufacture and Uses of Lime

The name "lime" is used for both calcium oxide (quicklime), and calcium hydroxide (slaked lime).

When limestone (calcium carbonate) is heated at about 1000 °C, it undergoes thermal decomposition; it loses carbon dioxide and turns into quicklime (calcium oxide).

Calcium carbonate   **calcium oxyde + carbon dioxyde**

$$CaCO_{3\ (s)} \longrightarrow CaO_{(s)} + CO_{2\ (g)}$$

The reaction is carried out in specially constructed lime kilns, (a kiln is a high temperature oven). Limestone is added at the top, and quicklime is removed from the bottom in a continuous process. The same reaction occurs in the blast furnace. Quicklime reacts with water to form slaked lime (calcium hydroxide). The reaction is highly exothermic.

Calcium oxyde + water   **calcium + hydroxyde**

$$CaO + H_2O \qquad\qquad Ca(OH)_2$$

Slaked lime is used to reduce the acidity of lakes and soils, see acid rain. It acts faster than powdered limestone but is more expensive. Slaked lime dissolves a little in water to form lime water.

Quicklime, more formally known as calcium oxide (CaO), is a caustic alkaline substance which is produced by heating limestone in specially designed kilns. There are a wide range of uses for quicklime, ranging from mortar to flux, and the substance has been used by humans for centuries. Many companies produce and sell quicklime, sometimes with specific chemical impurities which make it especially suitable to certain applications.

Humans have been aware of the steps needed to create quicklime for a very long time, and chemists believe that the generation of quicklime may be one of the oldest chemical reactions known to man. People have certainly been using quicklime all over the world for thousands of years; in Mesoamerica, for example, people treated corn with quicklime in a process known as nixtamalization, while in India quicklime was used in a mixture designed to waterproof boats. Today, quicklime is used in many industrial processes, some of which were developed hundreds of years ago.

Quicklime is also known as burnt lime, a reference to its manufacturing process, or simply lime. To make lime, limestone ($CaCO_3$) is broken up and shoveled into a kiln, which is heated to very high temperatures. The high temperatures release carbon dioxide (CO_2) from the stone, turning it into calcium oxide. After it is cooled, the calcium oxide can be ground into a power and packaged for sale.

Numerous things can impact the quality of the resulting quicklime, ranging from the temperature of the kilns to natural impurities in the stone. Because limestone is a natural product, it can sometimes be hard to control these impurities; as a result, companies which produce quicklime tend to test their product regularly, to ensure that is of high quality. The specialized kilns are also closely monitored to ensure that the limestone heats and cools at an appropriate rate.

Quicklime requires careful handling. As it sits, it can acquire CO_2 from the air, reverting to its original form. This means that it needs to be used quickly, especially once it has been mixed with water in a process known as slaking. Lime is also extremely caustic; it can burn the skin and cause other damages. When used responsibly, quicklime can be immensely useful for things like mixing strong mortar, acting as a flux in smelting, and treating wastewater, among many other things.

- There are a wide range of uses for quicklime, ranging from mortar to flux, and the substance has been used by humans for centuries. Many companies produce and sell quicklime, sometimes with specific chemical impurities which make it especially suitable to certain applications.
- The process for creating lime putty involves the combination of lime chalk or limestone fragments with water to produce a variety of different products. Lime putty or quicklime can be employed as a binding agent, a covering coat on a structure, or as one component in the creation of walkways or mosaics.

Hydrated Lime

- Type SA and NA are aerated limes, so their structure is more porous than their non-aerated counterparts. Hydrated limes are classified by the amount of water they retain and their maximum air content.
- This means that it needs air in order to be fully set. It can take hydrated lime plaster several months to completely dry, but it is desirable for construction due to ease of application.

Slaked Lime

- Lime wash is basically lime putty that has been watered down and can be easily applied to various surfaces. The slaked lime putty can be administered with the natural hue, or just about any color can be added to the limewashing before the actual application.
- NA and SA types are classified as having a maximum air content of 14 percent, and N and S types have seven percent air content. Also called slaked lime, hydrated lime is a good bonding agent and is water-tight. It is also strongly alkaline, having a pH of 12.

The clay content of limestone determines the quality of lime.

Percentage of clay less than 5%	fat lime
From 5 to 12% of clay	burned or quicklime
From 12 to 20% of clay	hydraulic lime

The hydraulic lime hardens slowly under water.

From 20 to clay 25%	slow hardening Portland cement
From 25 to clay 30%	quick hardening Portland cement

In chemical industry

a) for the manufacture of CO_2
b) for the manufacture of CaC_2, carburizes calcium, basic commodity of acetylene and its derivatives.
c) for the refining of sugar
d) for the dephosphorisation of steel in the Thomas process.
e) In lithography

Pozzolanic Reaction

This reaction is known since antiquity by the Romans to manufacture binders for their architectural works. The pozzolanic reaction is a phenomenon which transforms, at ordinary temperatures and in a reasonable time, the mixture pozzolanas, lime and water, in a hard and compact material. These pozzolanas can be natural or artificial.

This reaction can occur as well with natural pozzolanas as artificial. Indeed, one regards as pozzolanas any siliceous material or silico - aluminous which does not have, by itself, no flexible virtue, but which reacts chemically in the presence of water and of lime, to form products with the binding properties.

Thus, the natural products that are the volcanites, the diatomites and zeolites are gathered under the same term of "pozzolanas" as well as artificial materials (fly-ashes, waste of tiles, brick and forming refractory of the broken tiles,…), although they have very diverse origins. That is due to their properties, known as "pozzolanic". The pozzolanic activity results from a reaction with the products of the alkaline attack of silicates.

This reaction can be written:

Silica/alumina + lime + water hydrated calcic silicate/hydrated calcic aluminate

To observe this reaction, one needs that silica and alumina are very mobilizable, such as for example in an amorphous structure and more particularly of acid glasses. The pozzolanic activity of the burnt clays, amorphous, and of the fly-ashes, primarily vitreous, is related to this characteristic.

The products formed following this reaction, carried out at ordinary temperature are primarily hydrated calcic silicates (H.C.S.) of general formula $(CaO) (0.6-0.9) (SiO_2) (1-\beta) (Al_2O_3) \beta (H_2O) Z$, being represented in the form of filaments or of freezing with low crystallinity, and aluminates will tetra calcic hydrated (C4AH13) hexagonal.

The pozzolanic activity consists:

• in an alkaline attack superficial of acid silico-aluminous minerals by a solution saturated with lime,
• in the combination of the ions resulting from this attack with lime presents in the solution.

The formed products result similar to those from the hydration of the clinker and the granulated slag. They show interesting mechanical characteristics. Thus, thanks to the coalescence of the filaments of HSC and to the junction of encrusting of hydrates, recovering the mineral grains, these newly formed hydrated products have a mechanical resistance, an important compactness and durability.

References

"The Cement Industry 1796-1914: A History," by A. J. Francis, 1977.

P. C. Hewlett (Ed)*Lea's Chemistry of Cement and Concrete: 4th Ed*, Arnold, 1998, ISBN 0-340-56589-6, Chapter 1.

Housing Prototypes: Page Street.

Epidemiologic Notes and Reports Sulfur Dioxide Exposure in Portland Cement Plants.

http://www.azdeq.gov/function/news/2003/jan.html

CemNet.com | The latest cement news and information.

Toward a Sustainable Cement Industry: Environment, *Health and Safety Performance Improvement.* "As a generalization, probably 50% of all industrial byproducts have potential as raw materials for the manufacture of Portland cement."

Kosmatka, S.H.; Panarese, W.C. (1988). Design and Control of Concrete Mixtures. Skokie, IL, USA: Portland Cement Association, p. 15. ISBN 0-89312-087-1.

Chapter 6

BRICKS FROM PAST TO PRESENT

A brick is a block of ceramic material used in masonry construction, laid using mortar. The Roman Constantine Basilica in Trier, Germany, built in the 4th century with fired bricks as audience hall for Constantine I

The oldest shaped bricks found date back to 7,500 B.C. They have been found in Çayönü, a place located in the upper Tigris area, and in south east Anatolia close to Diyarbakir. Other more recent findings, dated between 7,000 and 6,395 B.C., come from Jericho and Catal Hüyük. From archaeological evidence, the invention of the fired brick (as opposed to the considerably earlier sun-dried mud brick) is believed to have arisen in about the third millennium BC in the Middle East. Being much more resistant to cold and moist weather conditions, brick enabled the construction of permanent buildings in regions where the harsher climate precluded the use of mud bricks. Bricks have the added warmth benefit of slowly storing heat energy from the sun during the day and continuing to release heat for several hours after sunset. The Ancient Egyptians and the Indus Valley Civilization also used mudbrick extensively, as can be seen in the ruins of Buhen, Mohenjo-daro and Harappa, for example. In the Indus Valley Civilization all bricks corresponded to sizes in a perfect ratio of 4:2:1

The world's highest brick tower of St. Martin's Church (Landshut), completed in 1500.

In Sumerian times offerings of food and drink were presented to "the Bone god," who was "represented in the ritual by the first brick." More recently, mortar for the foundations of the Hagia Sophia in Istanbul was mixed with "a broth of barley and bark of elm" and sacred relics, accompanied by prayers, placed between every 12 bricks. The Romans made use of fired bricks, and the Roman legions, which operated mobile kilns, introduced bricks to many parts of the empire. Roman bricks are often stamped with the mark of the legion that supervised its production. The use of bricks in Southern and Western Germany, for example, can be traced back to traditions already described by the Roman architect Vitruvius. In pre-modern China, brick-making was the job of a lowly and unskilled artisan, but a kilnmaster was respected as a step above the latter. Early descriptions of the production process and glazing techniques used for bricks can be found in the Song Dynasty carpenter's manual *Yingzao Fashi*, published in 1103 by the government official Li Jie, who was put in charge of overseeing public works for the central government's construction agency. The historian Timothy Brook writes of the production process in Ming Dynasty China (aided with visual illustrations from the *Tiangong Kaiwu* encyclopedic text published in 1637):

The brickwork of Shebeli Tower in Iran displays 12th century craftsmanship

...the kilnmaster had to make sure that the temperature inside the kiln stayed at a level that caused the clay to shimmer with the color of molten gold or silver. He also had to know when to quench the kiln with water so as to produce the surface glaze. To anonymous laborers fell the less skilled stages of brick production: mixing clay and water, driving oxen over the mixture to trample it into a thick paste, scooping the paste into standardized wooden frames (to produce a brick roughly 42 centimeters long, 20 centimeters wide, and 10 centimeters thick), smoothing the surfaces with a wire-strung bow, removing them from the frames, printing the fronts and backs with stamps that indicated where the bricks came from and who made them, loading the kilns with fuel (likelier wood than coal), stacking the bricks in the kiln, removing them to cool while the kilns were still hot, and bundling them into pallets for transportation. It was hot, filthy work. The idea of signing one's name on one's work and signifying the place where the product was made—in this case, bricks—was nothing new to the Ming era and had little or nothing to do with vanity. As far back as the Qin Dynasty (221 BC–206 BC), the government required blacksmiths and weapon-makers to engrave their names onto weapons in order to trace the weapon back to them, lest their weapons should prove to be of a lower quality than the standard required by the government.

In the 12th century, bricks from Northern Italy were re-introduced to Northern Germany, where an independent tradition evolved. It culminated in the so-called brick Gothic, a reduced style of Gothic architecture that flourished in Northern Europe, especially in the regions around the Baltic Sea which are without natural rock resources. Brick Gothic buildings, which are built almost exclusively of bricks, are to be found in Denmark, Germany, Poland and Russia. During the Renaissance and the Baroque, visible brick walls were unpopular and the brickwork was often covered with plaster. It was only during the mid-18th century that visible brick walls regained some degree of popularity, as illustrated by the Dutch Quarter of Potsdam, for example.

The transport in bulk of building materials such as paper over long distances was rare before the age of canals, railways, roads and heavy goods vehicles. Before this time bricks were generally made as close as possible to their point of intended use. It has been estimated that in England in the eighteenth century carrying bricks by horse and cart for ten miles (16 km) over the poor roads then existing could more than double their price. Bricks were often used, even in areas where stone was available, for reasons of speed and economy. The buildings of the Industrial Revolution in Britain were largely constructed of brick and timber due to the unprecedented demand created. Again, during the building boom of the nineteenth century in the eastern seaboard cities of Boston and New York, for example, locally made bricks were often used in construction in preference to the brownstones of New Jersey and Connecticut for these reasons. The trend of building upwards for offices that emerged towards the end of the 19th century displaced brick in favor of cast and wrought iron and later steel and concrete. Some early 'skyscrapers' were made in masonry, and demonstrated the limitations of the material - for example, the Monadnock Building in Chicago (opened in 1896) is masonry and just sixteen stories high, the ground walls are almost 1.8 meters thick, clearly building any higher would lead to excessive loss of internal floor space on the lower floors. Brick was revived for high structures in the 1950s following work by the Swiss Federal Institute of Technology and the Building Research Establishment in Watford, UK. This method produced eighteen story structures with bearing walls no thicker than a single brick (150-225 mm). This potential has not been fully developed because of the ease and speed in building with other materials, in the late-20th century brick was confined to low- or medium-rise structures or as a thin decorative cladding over concrete-and-steel buildings or for internal non-loadbearing walls.

METHODS OF MANUFACTURE

Brick making starts at the beginning of the 20th century. Bricks may be made from clay, shale, soft slate, calcium silicate, concrete, or shaped from quarried stone. Clay is the most common material, with modern clay bricks formed in one of three processes - soft mud, dry press, or extruded. In 2007 a new type of brick was invented, based on fly ash, a by-product of coal power plants.

Mud Bricks

The soft mud method is the most common, as it is the most economical. It starts with the raw clay, preferably in a mix with 25-30% sand to reduce shrinkage. The clay is first ground and mixed with water to the desired consistency. The clay is then pressed into steel moulds with a hydraulic press. The shaped clay is then fired ("burned") at 900-1000 °C to achieve strength.

ADOBE

Adobe bricks are a natural building material made from sand, clay, water, and some kind of fibrous material (sticks, straw, dung), which is shaped into bricks using frames and dried in the sun. It is similar to cob and mudbrick. Adobe structures are extremely durable and account for the oldest extant buildings on the planet. In dry climates, compared to wooden buildings adobe buildings offer significant advantages due to their greater thermal mass. Buildings made of sun-dried earth are common in the Middle East, North Africa, south America, southwestern North America, and in Spain (usually in the Mudéjar style). Adobe had been in use by indigenous peoples of the Americas in the Southwestern United States, Mesoamerica, and the Andean region of South America for several thousand years, although often substantial amounts of stone are used in the walls of Pueblo buildings. This method of brickmaking was imported to Spain in the 16th century by Spaniards who had traveled to Mexico and Peru. A distinction is sometimes made between the smaller *adobes,* which are about the size of ordinary baked bricks, and the larger *adobines,* some of which are as much as from one to two yards (2 m) long.

Etymology

The word *adobe* has come to us over some 4000 years with little change in either pronunciation or meaning: the word can be traced from the Middle Egyptian (c. 2000 BC) word *dj-b-t* "mud [*i.e.*, sun-dried] brickEnglish borrowed the word from Spanish in the early 18th century.

In more modern English usage, the term "adobe" has come to include a style of architecture that is popular in the desert climates of North America, especially in New Mexico. (Compare with stucco).

Composition of Adobe

An adobe brick is made of clay mixed with water and an organic material such as straw or dung. The soil composition typically contains clay and sand. Straw is useful in binding the brick together and allowing the brick to dry evenly. Dung offers the same advantage and is also added to repel insects. The mixture is roughly half sand (50%), one-third clay (35%), and one-sixth straw (15%).

Adobe Bricks

Bricks are made in an open frame, 25 cm (10 inches) by 36 cm (14 inches) being a reasonable size, but any convenient size is acceptable. The mixture is molded by the frame, and then the frame is removed quickly. After drying a few hours, the bricks are turned on edge to finish drying. Slow drying in shade reduces cracking.

The same mixture to make bricks, without the straw, is used for mortar and often for plaster on interior and exterior walls. Some ancient cultures used lime-based cement for the plaster to protect against rain damage.

The brick's thickness is preferred partially due to its thermal capabilities, and partially due to the stability of a thicker brick versus a more standard size brick. Depending on the form that the mixture is pressed into, adobe can encompass nearly any shape or size, provided drying time is even and the mixture includes reinforcement for larger bricks. Reinforcement can include manure, straw, cement, rebar or wooden posts. Experience has shown that straw, cement, or manure added to a standard adobe mixture can all produce a strong brick. A general testing is done on the soil content first. To do so, a sample of the soil is mixed into a clear container with some water, creating an almost completely saturated liquid. After the jar is sealed the container is shaken vigorously for at least one minute. It is then allowed to sit on a flat surface until the soil sediment has either collected on the bottom or remained a blended liquid. If the sediment collects on the bottom, that indicates there is a high clay content and is good for adobe. If the mixture remains a liquid, then there is little clay in the soil and using it would yield weak bricks. The largest structure ever made from adobe (bricks) was the Bam Citadel, which suffered serious damage (up to 80%) by an earthquake on December 26, 2003. Other large adobe structures are the Huaca del Sol in Peru, with 100 million signed bricks, the ciudellas of Chan Chan and Tambo Colorado, both in Peru.

Thermal Properties

An adobe wall can serve as a significant heat reservoir. A south-facing (in the Northern Hemisphere) adobe wall may be left uninsulated to moderate heating and cooling. Ideally, it should be thick enough to remain cool on the inside during the heat of the day but thin enough to transfer heat through the wall during the evening. The exterior of such a wall can be covered with glass to increase heat collection. In a passive solar home, this is called a Trombe wall. Adobe has a relatively dense thermal mass, and is most useful in tropical climates. In temperate climates it is less effective to heat a structure this way due to heat leaching by the ground and walls.

Adobe Wall Construction

When building an adobe structure, the ground should be compressed because the weight of adobe bricks is significantly greater than a frame house and may cause cracking in the wall. The footing is dug and compressed once again. Footing depth depends on the region and its ground frost level. The footing and stem wall are commonly 24" and 14", much larger than a frame house because of the weight of the walls. Adobe bricks are laid by course. Each course is laid the whole length of the wall, overlapping at the corners on a layer of adobe mortar. Adobe walls usually never rise above 2 stories because they're load bearing and have low structural strength. When placing window and door openings, a lintel is placed on top of the opening to support the bricks above. Within the last courses of brick, bond beams are laid across the top of the bricks to provide a horizontal bearing plate for the roof to distribute the weight more evenly along the wall. To protect the interior and exterior adobe wall, finishes can be applied, such as mud plaster, whitewash or stucco. These finishes protect the adobe wall from water damage, but need to be reapplied periodically, or the walls can be finished with other nontraditional plasters providing longer protection.

Adobe Roof

The traditional adobe roof has been generally constructed using a mixture of soil/clay, water, sand, and other available organic materials. The mixture was then formed and pressed into wood forms producing rows of dried, earth bricks that would then be laid across a support structure of wood and plastered into place with more adobe. For a deeper understanding of adobe, one might examine a cob building. Cob, a close cousin to adobe, contains proportioned amounts of soil, clay, water, manure, and straw. This is blended, but not formed like adobe. Cob is spread and piled around a frame and allowed to air dry for several months before habitation. Adobe, then, can be described as dried bricks of cob, stacked and mortared together with more adobe mixture to create a thick wall and/or roof.

Roof Materials

Depending on the materials available, a roof can be assembled using lengths of wood or metal to create a frame work to begin layering adobe bricks. Depending on the thickness of the adobe bricks, the frame work has been performed using a steel framing and a layering of a metal fencing or wiring over the framework to allow an even load as masses of adobe are spread across the metal fencing like cob and allowed to air dry accordingly. This method was demonstrated with an adobe blend heavily impregnated with cement to allow even drying and prevent major cracking.

Traditional Adobe Roof

More traditional adobe roofs were often flatter than the familiar steeped roof as the native climate yielded more sun and heat than mass amounts of snow or rain that would find use in precipitous roofs. Moisture, however, is often foe to a composite of mud and organic matter, so the introduction of cement is often more common to help ward off any undue water damage. It is at this turn that sense is required before the construction of any adobe is begun, be sure that the location for such a structure is similar to the climate it naturally comes from, that is, a hot, arid climate. Cool and moist climates would do well with moisture precautions planned out.

Raising a Traditional Adobe Roof

To raise a flattened adobe roof, beams of wood or metal should be assembled and span the extent of the building. The ends of the beams should then be fixed to the tops of the walls using the builder's preferred choice of attachments. Taking into account the material the beams and walls are made from, choosing the attachments may prove difficult. In combination to the bricks and adobe mortar that are laid across the beams creates an even load-bearing pressure that can last for many years depending on attrition. Once the beams are laid across the building, it is then time to begin the placing of adobe bricks to create the roof. An adobe roof is often laid with bricks slightly larger in width to ensure a larger expanse is covered when placing the bricks onto the beams. This wider shape also provides the future homeowner with thermal protection enough to stabilize an even temperature through out the year. Following each individual brick should be a layer of adobe mortar, recommended to be at least an inch thick to make certain there is ample strength between the brick's edges and also to provide a relative moisture barrier during the seasons where the arid climate does produce rain.

Attributes

Adobe roofs can be inherently fire-proof, an attribute well received when the fireplace is kept lit during the cold nights, depending on the materials used. This feature leads the homeowner and builders to begin thinking about the installation of a chimney, a feat regarded as a necessity in any adobe building. The construction of the chimney can also greatly influence the construction of the roof supports, creating an extra need for care in choosing the right materials. An adobe chimney can be made from simple adobe bricks and stacked in similar fashion as the surrounding walls. Basically outline the location and perimeter of the hearth, minding the safety elements common to a fireplace, and begin to stack and mortar the walls with pre-made adobe bricks, cut to size.

RAIL KILNS

In modern brickworks, this is usually done in a continuously fired tunnel kiln, in which the bricks move slowly through the kiln on conveyors, rails, or kiln cars to achieve consistency for all bricks. The bricks often have added lime, ash, and organic matter to speed the burning.

Bull's Trench Kilns

In Pakistan and India, brick making is typically a manual process. The most common type of brick kiln in use there are Bull's Trench Kiln (BTK), based on a design developed by British engineer W. Bull in the late 1800s. An oval or circular trench, 6-9 meters wide, 2-2.5 meters deep, and 100-150 meters in circumference, is dug in a suitable location. A tall exhaust chimney is constructed in the center. Half or more of the trench is filled with "green" (unfired) bricks which are stacked in an open lattice pattern to allow airflow. The lattice is capped with a roofing layer of finished brick. In operation, new green bricks, along with roofing bricks, are stacked at one end of the brick pile; cooled finished bricks are removed from the other end for transport. In the middle the brickworkers create a firing zone by dropping fuel (coal, wood, oil, debris, etc) through access holes in the roof above the trench. The advantage of the BTK design is a much greater energy efficiency compared with clamp or scove kilns. Sheet metal or boards are used to route the airflow through the brick lattice so that fresh air flows first through the recently burned bricks, heating the air, then through the active burning zone. The air continues through the green brick zone (pre-heating and drying them), and finally out the chimney where the rising gases create suction which pulls air through the system. The reuse of heated air yields a considerable savings in fuel cost. As with the rail process above, the BTK process is continuous. A half dozen laborers working around the clock can fire approximately 15,000-25,000 bricks a day. Unlike the rail process, in the BTK process the bricks do not move. Instead, the locations at which the bricks are loaded, fired, and unloaded gradually rotate through the trench.

Dry Pressed Bricks

The dry press method is similar to mud brick but starts with a much thicker clay mix, so it forms more accurate, sharper-edged bricks. The greater force in pressing and the longer burn make this method more expensive.

Extruded Bricks

In extruded bricks the clay mix is 20-25% water, this is forced through a die to create a long cable of material of the proper width and depth. This is then cut into bricks of the desired length by a wall of wires. Most structural bricks are made by this method, as hard dense bricks result, and holes or other perforations can be produced by the die. The introduction of holes reduces the needed volume of clay through the whole process, with the consequent reduction in cost. The bricks are lighter and easier to handle, and have thermal properties different from solid bricks. The cut bricks are hardened by drying for between 20 and 40 hours at 50-150 °C before being fired. The heat for drying is often waste heat from the kiln.

Calcium Silicate Bricks

The raw materials for calcium silicate bricks include lime mixed with quartz, crushed flint or crushed siliceous rock together with mineral colorants. The materials are mixed and left until the lime is completely hydrated, the mixture is then pressed into moulds and cured in an autoclave for two or three hours to speed the chemical hardening. The finished bricks are very accurate and uniform, although the sharp arrises need careful handling to avoid damage to brick (and brick-layer). The bricks can be made in a variety of colours, white is common but a wide range of "pastel" shades can be achieved.

Fly Ash Bricks

In May 2007, Haoxaing Fei, a retired civil engineer, announced that he had invented a new brick composed of fly ash and water compressed at 4,000 psi (27,939 kPa) for two weeks. Owing to the high concentration of calcium oxide in fly ash, the brick is considered "self-cementing". The brick is toughened using an air entrainment agent, which traps microscopic bubbles inside the brick so that it resists penetration by water, allowing it to withstand up to 100 freeze-thaw cycles. Since the manufacturing method uses a waste by-product rather than clay, and solidification takes place under pressure rather than heat, it has several important environmental benefits. It saves energy, reduces mercury pollution, alleviates the need for landfill disposal of fly ash, and costs 20% less than traditional clay brick manufacture. Liu intends to license his technology to manufacturers in 2008.

Influence on Fired Colour

The fired colour of clay bricks is significantly influenced by the chemical and mineral content of raw materials, the firing temperature and the atmosphere in the kiln. For example pink coloured bricks are the result of a high iron content, white or yellow bricks have a higher lime content.

Most bricks burn to various red hues, if the temperature is increased the colour moves through dark red, purple and then to brown or grey at around 1300 °C. Calcium silicate bricks have a wider range of shades and colours, depending on the colorants used. Bricks formed from concrete are usually termed blocks, and are typically pale grey in colour. They are made from a dry, small aggregate concrete which is formed in steel moulds by vibration and compaction in either an "egglayer" or static machine. The finished blocks are cured rather than fired using low-pressure steam. Concrete blocks are manufactured in a much wider range of shapes and sizes than clay bricks and are also available with a wider range of face treatments - a number of which are to simulate the appearance of clay bricks.

An impervious and ornamental surface may be laid on brick either by salt glazing, in which salt is added during the burning process, or by the use of a "slip," which is a glaze material into which the bricks are dipped. Subsequent reheating in the kiln fuses the slip into a glazed surface integral with the brick base.

Natural stone bricks are of limited modern utility, due to their enormous comparative mass, the consequent foundation needs, and the time-consuming and skilled labour needed in their construction and laying. They are very durable and considered more handsome than clay bricks. Only a few stones are suitable for bricks. Common materials are granite, limestone and sandstone. Other stones may be used (e.g. marble, slate, quartzite, etc.) but these tend to be limited to a particular locality.

Optimal Dimensions, Characteristics and Strength

Loose Bricks

For efficient handling and laying bricks must be small enough and light enough to be picked up by the bricklayer using one hand (leaving the other hand free for the trowel). Bricks are usually laid flat and as a result the effective limit on the width of a brick is set by the distance which can conveniently be spanned between the thumb and fingers of one hand, normally about four inches (about 100 mm).

In most cases, the length of a brick is about twice its width, about eight inches (about 200 mm) or slightly more. This allows bricks to be laid *bonded* in a structure to increase its stability and strength.

The wall is built using alternating courses of *stretchers*, bricks laid longways and *headers*, bricks laid crossways. The headers tie the wall together over its width. The correct brick for a job can be picked from a choice of color, surface texture, density, weight, absorption and pore structure, thermal characteristics, thermal and moisture movement, and fire resistance.

In England, the length and the width of the common brick has remained fairly constant over the centuries, but the depth has varied from about two inches (about 51 mm) or smaller in earlier times to about two and a half inches (about 64 mm) more recently.

In the United States, modern bricks are usually about 8 × 4 × 2.25 inches (203 × 102 × 57 mm). In the United Kingdom, the usual ("work") size of a modern brick is 215 × 102.5 × 65 mm (about 8.5 × 4 × 2.5 inches), which, with a nominal 10 mm mortar joint, forms a "coordinating" or fitted size of 225 × 112.5 × 75 mm, for a ratio of 6:3:2.

Face brick ("house brick") sizes, from small to large

Standard	Imperial	Metric
United States	8 × 4 × 2¼ inches	203 × 102 × 57 mm
United Kingdom	8½ × 4 × 2½ inches	215 × 102.5 × 65 mm
South Africa	8¾ × 4 × 3 inches	222 × 106 × 73 mm
Australia	9 × 4⅓ × 3 inches	230 × 110 × 76 mm

Some brickmakers create innovative sizes and shapes for bricks used for plastering (and therefore not visible) where their inherent mechanical properties are more important than the visual ones. These bricks are usually slightly larger, but not as large as blocks and offer the following advantages:

- A slightly larger brick requires less mortar and handling (fewer bricks) which reduces cost
- Ribbed exterior aids plastering
- More complex interior cavities allow improves insulation, while maintaining strength.

Blocks have a much greater range of sizes. Standard coordinating sizes in length and height (in mm) include 400×200, 450×150, 450×200, 450×225, 450×300, 600×150, 600×200, and 600×225; depths (work size, mm) include 60, 75, 90, 100, 115, 140, 150, 190, 200, 225, and 250. They are usable across this range as they are lighter than clay bricks. The density of solid clay bricks is around 2,000 kg/m³: this is reduced by frogging, hollow bricks, etc.; but aerated autoclaved concrete, even as a solid brick, can have densities in the range of 450–850 kg/m³.

Bricks may also be classified as *solid* (less than 25% perforations by volume, although the brick may be "frogged," having indentations on one of the longer faces), *perforated* (containing a pattern of small holes through the brick removing no more than 25% of the volume), *cellular* (containing a pattern of holes removing more than 20% of the volume, but closed on one face), or *hollow* (containing a pattern of large holes removing more than 25% of the brick's volume). Blocks may be solid, cellular or hollow

The term "frog" for the indentation on one bed of the brick is a word that often excites curiosity as to its origin. The most likely explanation is that brickmakers also call the block that is placed in the mould to form the indentation a frog. Modern brickmakers usually use plastic frogs but in the past they were made of wood. When these are wet and have clay on them they resemble the amphibious kind of frog and this is where they got their name. Over time this term also came to refer to the indentation left by them.[*Matthews 2006*]

The compressive strength of bricks produced in the United States ranges from about 1000 lbf/in² to 15,000 lbf/in² (7 to 105 MPa or N/mm²), varying according to the use to which the brick are to be put. In England clay bricks can have strengths of up to 100 MPa, although a common house brick is likely to show a range of 20–40 MPa.

USE

In the early 1900s, most of the streets in the city of Grand Rapids, Michigan were paved with brick. Today, there are only about 20 blocks of brick paved streets remaining (totaling less than 0.5 percent of all the streets in the city limits).

Bricks are used for building and pavement. In the USA, brick pavement was found incapable of withstanding heavy traffic, but it is coming back into use as a method of traffic calming or as a decorative surface in pedestrian precincts.

Bricks are also used in the metallurgy and glass industries for lining furnaces. They have various uses, especially refractory bricks such as silica, magnesia, chamotte and neutral (chromomagnesite) refractory bricks. This type of brick must have good thermal shock resistance, refractoriness under load, high melting point, and satisfactory porosity. There is a large refractory brick industry, especially in the United Kingdom, Japan and the U.S.A..

In the United Kingdom, bricks have been used in construction for centuries. Until recently, many houses were built almost entirely from red bricks. This use is particularly common in areas of northern England and some outskirts of London, where rows of terraced houses were rapidly and cheaply built to house local workers. These houses have survived to the present day. Although many houses in the UK are now built using a mixture of concrete blocks and other materials, many houses are skinned with a layer of bricks on the outside for aesthetic appeal.

REFERENCES

Brook, Timothy. (1998). *The Confusions of Pleasure: Commerce and Culture in Ming China*. Berkeley: University of California Press. ISBN 0-520-22154-0.

Campbell, James W. P., and Will Pryce. 2003. *Brick: a world history*. London ; New York: Thames and Hudson.

M.Kornmann and CTTB, Clay bricks and roof tiles, manufacturing and properties, Lasim (Paris) 2007 ISBN 2-9517765-6-X.

Chapter 7

SOME MODERN PROCESSES
TO MAKE BRICKS

MORTARS WITH NATURAL TUFFS FILLERS:
DURABILITY IN AGGRESSIVE ENVIRONNEMENT

Abstract

Huge quantities of volcanic tuffs are found in Mako area (western Senegal). The high content of SiO_2 is responsible of the pozzolanic properties of the tuffs when mixed to Portland cement. However, tuffs alterations could limit the pozzolanic effect: for example, the formation of clay decreases this effect but decrease also the porosity.

The aim of this study is to highlight the benefits of this admixture when mortars where exposed in different sulphatic solutions. In sulphuric acid, specimen surfaces are covered with a superficial formation of gypsum and with amonium sulphate solution only few needles of gypsum appears. As it is shown by SEM observations and X-ray diffraction, deeper alterations decrease when cement is partially substituted with tuff.

Keywords: durability, SEM, X-ray diffraction, sulphatic solutions, volcanic tuffs

Introduction

Volcanic tuffs, like amorphous silica or silica with poorly crystallised phases (silica fume, fly ash, natural materials: diatomite), exhibit pozzolanic properties in mortars when cement is partially substituted (Papadakis and Tsimas, 2002). In a similar study, volcanic tuffs from Galatean Province of Central Turkey where used by Turkmenoglu et al, 2002. Their investigations with petrographic observations highlight the influence of different parameters (proportion of glass, clay minerals) in the pozzolanic effects. Workability was modified when cement is partially substituted with tuff. Investigations of Uzal and Turanli, 2003 showed the influence of grinding times, particle size distribution and the super-plasticizer influence to maintain good workability. In this case, the compressive strength is lowered when tuff is added and alkali silica expansion is reduced. Compression and flexural strengths of concrete containing tuff and cured in hot weather condition decrease in comparison with normal concrete (Ujhelyi et al, 1991).

The use of tuffs from Mako Province in western Senegal was tested in our laboratory by Coatanlem et al, 2004. However, tuff alterations could limit the pozzolanic effect: for example, the formation of clay decreases this effect. This paper highlights the deteriorations of mortars with tuffs subjected to sulphate environments. The action of sulphuric acid and ammonium sulphate solutions with 3 concentrations is considered. Similarly, normal mortars where tested by Jauberthie and Rendell, 2003 and by Rendell et al, 2002.

Samples Preparation

Mixture Compounds

The samples used in the test program were 4cm x 4cm x 16cm prisms based on a standard cement mortar mix. The mortar used in the test is a normal mortar conforming to EN196. The aggregate type being standard sand conforming to ISO 679 and the method of manufacture was carried out in accordance to EN 196. The mixture details were the following: aggregate-to-cement ratio was 3.0 and the water-to-cement ratio was 0.5. The cement type used is a normal portland cement – CEM 1 52.5 CP2 manufactured in SAINT-PIERRE-LA-COUR (France). An analysis of the cement type is summarized in Table 1. In the second series of samples, the volcanic tuff, ground below 50μm, replaced partially the cement. An analysis of the volcanic tuff is summarized in Table 2. SEM examinations of this tuff are shown in figure 1.

The X-Ray diffraction analysis (figure 2) reveals the presence of illite and kaolin: 2 clay minerals, as well as quartz and goethite.

Table 1. Oxide analysis of the cement used in the tests

Ins	SiO_2	Al_2O_3	Fe_2O_3	CaO	MgO
0.26	20.15	5.18	2.76	65.13	0.69
K_2O	Na_2O	SO_3	P.F.	CaOfree	
0.99	0.17	2.85	1.51	1.31	

Table 2. Oxide analysis of the tuff used in the tests

SiO_2	Al_2O_3	Fe_2O_3	MgO	K_2O	Na_2O
82.54	7.18	9.08	0.27	0.58	0.34

Figure 1. SEM micrographs of the volcanic tuff.

In the investigation, the two series of samples were produced and subjected to identical regime of exposure. Mortar composition, flexural and compression strengths of samples MO (normal mortar) and MT (mortar with tuff) measured after 2 years (stored in an air conditioned room 20°C, 50%RH) are summarised in Table 3. A pozzolanic effect of tuff is expected. But, our tuff is altered and which explain a content of clay (illite and kaolinite). The tuff content increase induces lower strength due to the competition between pozzolanic effect and fillers effect.

Table 3. Mortar properties of reference samples after 2 years

	Composition	by weight (g)	Flexural strength (MPa)	Compressive strength (MPa)
MO	Sand	2700	11.2	66.5
	Cement	900		
	Tuff	0		
	Water	450		
MT	Sand	2700	7.2	42.5
	Cement	630		
	Tuff	270		
	Water	450		

Exposure Tests in Sulphate Environment

After 2 years storage in an air conditioned room (20°C, 50%RH), the following stage in the program consisted in the comparison of the characteristics of the mortars with and without tuff throughout 2 months exposure in sulphate solutions (Figure 2). Six curing environments were used in the test:

- Stored in sulphuric acid, 0.1M, 0.25M, 0.5M,
- Stored in ammonium sulphate, 0.1M, 0.25M, and 0.5M.

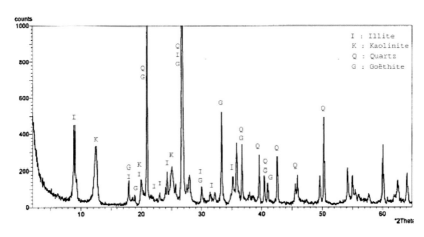

Figure 2. XRD pattern of tuff (λ k$_\alpha$ Cu , filtered).

Figure 3. Exposure of samples in sulphate solutions (six curing environments).

Table 4. Evolution of the section of the samples v/s solutions (H_2SO_4)

	Stored solution	Size (mm)
MO	H_2SO_4 0.1 M	40.5
	H_2SO_4 0.25M	38.9
	H_2SO_4 0.5 M	36.5
MT	H_2SO_4 0.1 M	40.5
	H_2SO_4 0.25M	40.8
	H_2SO_4 0.5 M	41.8

Tests Results

Sulphuric Acid

It was noted that during the test period, the samples in the sulphuric acid solution were covered in a soft white deposit. The degree of samples surface deterioration was very evident (figure 4) as it can be seen in table 4 were the size changes of the samples are reported. The section of normal mortar decrease largely when the sample is in the acidic solution. When the concentration increases, the size decreases.

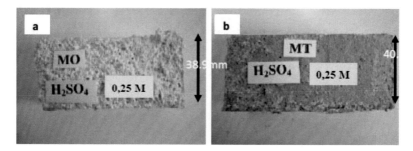

Figure 4. Samples stored in H_2SO_4 (0.25M).

Table 5. Evolution of the section of the samples v/s solutions ($(NH_4)_2SO_4$)

	Stored solution	Size (mm)
MO	$(NH_4)_2SO_4$ 0.1 M	40.1
	$(NH_4)_2SO_4$ 0.25 M	40.2
	$(NH_4)_2SO_4$ 0.5 M	40.4
MT	$(NH_4)_2SO_4$ 0.1 M	40.1
	$(NH_4)_2SO_4$ 0.25 M	40.1
	$(NH_4)_2SO_4$ 0.5 M	40.1

Ammonium Sulphate

Lee (1970) suggests that ammonium sulphate is the most destructive of the sulphate salts: a 5% solution is able to causing 3.8% linear expansion in a cement mortar after 12 weeks. He proposes that the expansion may be due to the formation of an expansive double salt $CaSO_4.(NH_4)$ $2SO_4.H_2O$. Considering the general reaction between an ammonium salt and lime it can be seen that one would expect the formation of $CaSO_4.2H_2O$ gypsum and a release of ammoniac; this gas liberation being common to most ammonium salts. It has been observed that during the period of immersion of a PC CEM I mortar in ammonium sulphate solution there is the formation of needle like crystals over the surface of the concrete, these have been identified as gypsum (Rendell and Jauberthie, 1999; Rendell et al., 2000). Samples stored in ammonium sulphate present some needle like crystal at the surface (figure 5) and the samples sizes were not significantly modified (Table 5).

Analysis of the Product Resulting from the Attack

After the acidic attack, the solid residue is separated by filtration. After dehydration at 25°C, the X-Ray diffraction analysis shows that the attack by the sulphuric acid transforms the hardened cement paste into gypsum similarly with normal mortar (figure 6) and tuff + cement mortar (figure 7). Naturally, the quartz is found in the 2 samples, and clay minerals traces (illite and kaolin) were detected. Those residues have to be dried at a temperature lower than 40°C to avoid the transformation of gypsum into hemi hydrate.

Figure 5. Samples stored in $(NH_4)_2SO_4$ (0.25M).

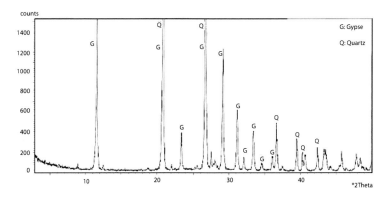

Figure 6. XRD pattern of reference sample MO stored in H_2SO_4 (0.25M) (λ k_α Cu , filtered).

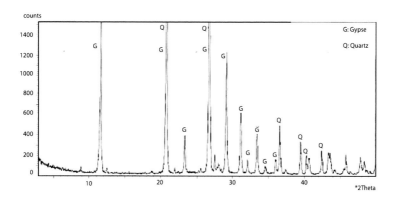

Figure 7. XRD pattern of sample with tuff MT stored in H2SO4 (0.25M) (λ kα Cu , filtered).

SEM Observations

SEM observations (figure 8) of the sample of normal mortar stored in sulfuric acid (H_2SO_4 0.25M) confirm the X-Ray diffraction results. A formation of scale (figure 8 detail (0)) on the sample surfaces is noted. Such scale induces a penetration of acidic solution deeper in the section. The falling of scales explains the section decreases observed in figure 3. Large prismatic crystals of gypsum cover the surface and fill the pores. Behind this layer, an ettringite front appears. Fine needle like crystals are present to 1.5 mm depth. Portlandite layer are not observed in this area. At 2 mm depth, portlandite is slightly altered (figure 8, detail (5)). Deeper in, traditional hydrated mineral phases of classic Portland Cement are observed.

SEM observations of the sample of mortar containing tuff and stored in sulfuric acid (H_2SO_4 0.25M) are presented on figure 9. Same gypsum mineralization is observed at the sample surface. However, no scale appears (figure 9 detail (3)). At 0.5 mm depth, the portlandite layers are detected and at 1.3 mm the portlandite is less altered. The formation of new mineral phases without spalling explain the expansion of samples (table 4).

The tuff addition induces two type of effect. A pozzolanic effect is due to silica. This effect is limited because of natural tuff alteration. Induced clays bring a filler effect, not interesting regarding strength but reducing the global pore size. Combination of those two effects leads to delay of the sulphatic attack.

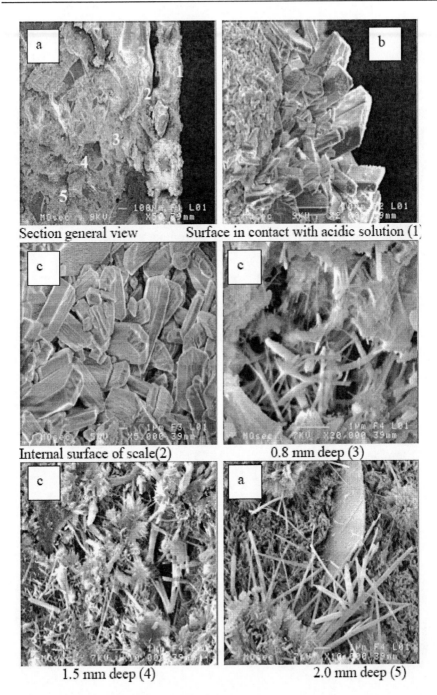

Figure 8. MO Samples stored in H_2SO_4 (0.25 M).

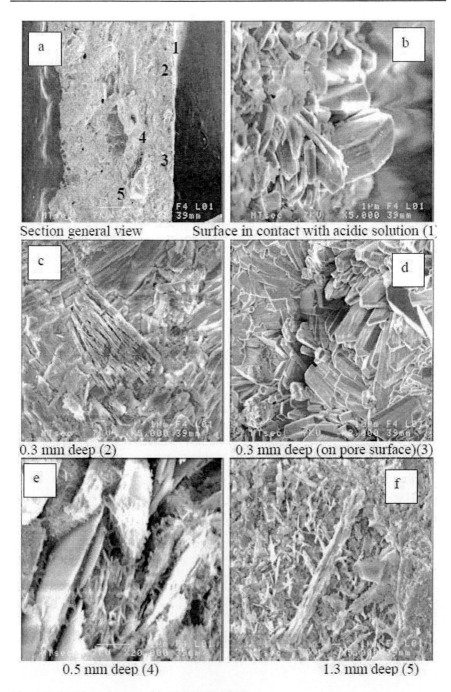

Figure 9. MT Samples stored in H_2SO_4 (0.25 M).

Valorization of Tuffs: Setting Forms by Extrusion

From this study, it appeared interesting to improve qualities of the mortars by increasing the compactness of the mixtures. The setting forms by extrusion are then considered. This traditional technique for the production of argillaceous products is generally not easily exploitable in the presence of a granular mixture rubbing. Consequently, in the present study we focused our attention on the extrusion of pastes of tuffs stabilized with cement. The compositions are such the incorporated cement replaces the tuffs in proportions ranging between 4 and 10%, current proportioning while binding to carry out "earths concrete". However only the results obtained for 6% are presented here. The ponderal reports/ratios Water/Dry Masse (W/DM) tested on the rheological level are 25, 30, 35 and 40%. Preliminary tests of identification of the rheological behaviour are carried out with the aim of evaluating the extrudability of the tuffs: measurements of the threshold of shearing with the rotating auger and tests of simple compression with striated plates, carried out on mixtures of tuffs and water. The results of the shear tests are presented in table 6. Similarly to work of Toutou et al. (2005) on the mortars containing cement, the results of the compression tests (Figure 6) reveal a primarily plastic rubbing character. For a pure, famous plastic material extrudable, the recording of $F^* = F^*h/ (\pi^*R^3)$ (where F is the compressive force, R the ray of the plates and H the height of the sample), is a line of null slope.

Table 6. Thresholds of flow obtained by direct shearing according to the water content of the pastes of tuffs

Water content (%)	30	35	40
Thresholds of flow (kPa)	33.2	15.4	4.4

The speed of test modifies the answer of the test. This is explained by a phenomenon of drainage and draining of the central part of the sample which occurs for the lowest speed. The measurement of water content of the samples after compression made it possible to confirm this conclusion. Such drainage is penalizing with respect to the extrudability.

The apparently prevalent plastic character obtained for E/MS = 40% (Curves (1) of figure 12)) indicates a good aptitude for extrusion. However, the threshold of flow highlighted, table 6, proves too weak. The extrudat, at exit of die, can present defects by indentation or deformation under actual weight during the handling of the products, for example. Also, we agree to retain for the continuation of work the mixture W /DM = 35% whose value of the threshold of shearing is of 15,4 kPa, acceptable value to ensure the extrudability of argillaceous products Toutou et al. (2005).

Tests of extrusion are then considered on a paste of tuffs stabilized with cement: cement/tuffs = 6%; W/DM = 35%. In addition, an addition of 0.2% of the water mass of plasticizing (Sika, plastiment 22S) is incorporated in this mixture to limit the effects of reinforcement of the rubbing character of the mixture induced by the introduction of cement. The tests of extrusion are carried out using a laboratory extrusion machine of single-screw (Figure 10). The die of working is square of 4 cm side. The number of revolutions of the screws is 20tr/min while the number of revolutions of the supplying screw is 22 tr/min.

The extrudats obtained have a good aspect of surface and are free from wrenching in particular on the level of the edges (figure 11). A batch of 16 cm length test-tubes is manufactured. At exit of die, the thresholds of shearing measured with the rotating auger on the sides of the extrudats present a median value of 50,8kPa + /-4 kPa.

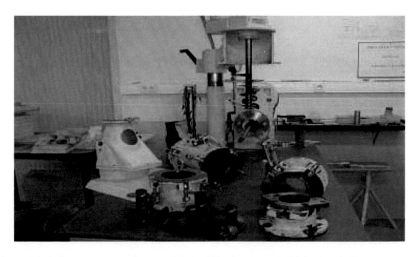

Figure 10. Laboratory extrusion machine of single-screw used in the study.

Figure 11. Batch of 16 cm length test-tubes is manufactured.

The pondered stabilization of the test-tubes is obtained after 12 days with storage at the ambient air. The average apparent bulk density is 1630kg/m³. At the end of 28 and 90 days curing, we determined the ultimate resistances of rupture in compression and traction (by a 3 points inflection test). The results are presented on figure 13. Similarly to the mechanical tests on the mortars of tuffs, the mechanical resistances are notably increased after 90 days curing (25% in compression, 50% in traction).

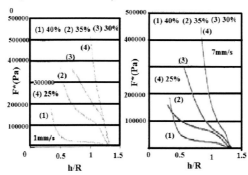

Figure 12. Comparison of tuffs paste: Influence of different water content and the speed of compression for different values of W/DM.

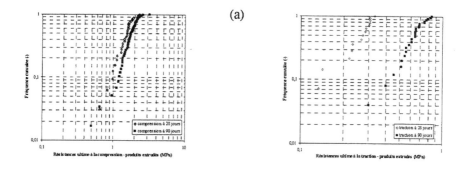

Figure 13. Functions of distribution of the ultimate resistances of rupture at 28 and 90days: (a) in compression and (b) in traction.

However, these resistances are low while remaining traditional for an "earth concrete" stabilized with cement. The functions of distribution of these resistances present a Gaussian character what tends to show certain homogeneity of the manufacturing process.

Finally the possibility of extruding this type of material is shown. The end product presents a good aspect and the first results obtained are encouraging even if the physical properties and mechanical must be improved. In addition, it is now advisable to consider the realization of extrusion with the mortars more proportioned out of cement.

Conclusion

Huge quantities of volcanic tuffs are found in Mako area (western Senegal). The high content of SiO_2 is responsible of the pozzolanic properties of the tuffs when mixed with Portland Cement. However, tuffs alterations could limit the pozzolanic effect: for example, the formation of clay decreases this effect. Mechanical tests are displayed on Portland Cement + tuffs and Portland Cement samples: 30% of tuff/cement substitution leads to compressive strength reduced by 35%.

When mortars were exposed to different sulphatic solutions, tuff increased chemical resistance. In sulphuric acid, specimen surface are covered with a superficial formation of gypsum. The spalling of this layer is reduced when tuff is present in the mortar. With ammonium sulphate, only few needles of gypsum grow at the surface of sample without tuff. In this case, smalls cracks appears at the surface when samples were washed.

References

Papadakis, V. G., Tsimas, S., "Supplementary cementing materials in concrete: Part 1. Efficiency and design"; *Cement and Concrete Research*, v 32, n 10, October 2002, pp. 1525-1532.

Turkmenoglu, Asumant, G. , Tankut, A. "Use of tuffs from central Turkey in pozzolanic cements : Assessment of their petrological properties"; *Cement and Concrete Research*, v 32, n 4, April 2002, pp. 629-637.

Uzal, B., Turanli, L. "Studies on blended cements containing a high volume of natural pozzolans"; *Cement and Concrete Research*, v 33, n 11, November 2003, pp. 1777-1781.

Ujhelyi, Janos E., Ibrahim, Ahmad J., "Hot weather concreting with hydraulic additives" ; *Cement and Concrete Research*, v 21, n 2-3,March-May 1991, pp. 345-354.

Coatanlem, P., Jauberthie, R., Diop,B. ; "Pouzzolanicité des tuffs volcaniques acides du Sénégal oriental" , Congrés AUGC, 3 and 4 juin, Marne-la-Vallée 2004.

Jauberthie, R., Rendell, F. ; "Physicochemical study of the alteration of concrete exposed to ammonium salts."; *Cement and Concrete Research*, v 33, n 1, 2003 , pp. 85-91.

Rendell, F., Jauberthie, R., Grantham, M. "Deteriorated concrete"; Ed. Thomas Telford Publishing 2002.

Lee, F.M., The chemistry of cement and concrete, Edward Arnold, 3rd Edition, London, 1970.

Rendell F. Jauberthie R. The deterioration of mortar in sulphate environments. *Construction and Building Materials*, 1999, 13, pp. 321 – 327.

Rendell F. Jauberthie R. Camps J.P. The effect of surface gypsum deposits on the durability of cementitious mortars under sulphate attack. Concrete Science and Engineering. RILEM, April 2000, Vol 2, pp. 231-244.

MODERN BRICKMAKING PROCESS

A. Calcium Silicate Bricks

A Case Study

Activation of the pouzzolanic reaction with phosphogypsum : application to the volcanic tuffs of eastern Senegal (west Africa)

Abstract

In Senegal, waste called phosphogypsum (Prayon, 1996) results from the manufacture of the phosphoric acid from phosphate. They are laid out in heap whose quality and quantity remain unspecified. The evaluation of the phosphogypsum stock shows a quantity available estimated at 166 million m^3.

From the environmental point of view, the scrubbing of these heaps by rainwater during the rain season makes the neighbouring soils acid. The area of Mako (Senegal Eastern) possess significant amount of tender and pulverulent pumice tuffs coming from the calco-alkali volcanism dating from the Birimien (approximately 2 billion years) era.

These pumice tuffs are acid (with nearly 70% of SiO_2). This high silica and aluminium content (13%) confers on the tuffs-cement mixtures pozzolanic properties (Coatanlem, 2004).

Treated with lime (2.5%), the tuffs make it possible to obtain bricks of high resistances (3 MPa) after 72 hours of conservation at 80°C. The phosphogypsum accelerates the pozzolanic reaction considerably and induces a gain in resistance of almost 60% allowing the use of the tuffs in road geotechnics and for construction.

Keywords: pumice tuffs, pozzolanic activity, treatment, lime, road, phosphogypsum, habitat

Introduction

The tender non-cohesive tuffs dating from birimien (approximately 2 billion years) of Mako are very old weathered and devitrified pyroclastites (N'Diaye and al. (2003). They form broad outcrops in Eastern Senegal and present considerable reserves. They constitute the main part of the sub grades in this area. They show poor geotechnical characteristics. Thanks to their smoothness and their tenderness, these materials suit well to a treatment by binders. The work presented here gives a progress report of the treatment with quicklime (CaO) of these tuffs and the incidence of the phosphogypsum addition on the pozzolanic reaction.

Table 7. Chemical composition of tuffs used and reactive phase

SiO2	Al2O3	Fe2O3	CaO	MgO	K2O	Na2O	TiO2	Mn2O3	PF	reactive silica
69.7	13.3	10.1	0.1	0.1	1.3	0.6	0.8	0.1	4	8.9

Materials and Methods of Study

The tuffs used come from Eastern Senegal (Bafoundou). Due to their tender character, they are easily pulverized with the jaw crusher. Then, the refusal to 250 μm is taken again to the ring crusher. It is the passing to 250 μm which was used in the formulations. The chemical analysis of the tuffs is given by table 7. It is an acid tuff (SiO_2 > 66%), rich in iron and alumina. The difference between the silica and lime content higher than 34% traduced a great vitreous fraction in the tuffs.

To determine approximately the amplitude of the tenor of lime to test, we have used the method proposed by Pichon (1992). This method is based on the determination of Potential combined lime (PCL). It's the optimal combined lime by unit mass of material.

Concerning the physical properties, they are summarized on table 2. The density of tuffs is superior to the one of silica (2.65) due its high iron content. It's a material which possess a considerable specific surface which twice those of current Portland cement. This fineness (figure 14) of the material is confirmed by laser particle size which gives an average grain size around 16 μm. According to Casagrande, the tuffs could be classified as slightly plastic silt.

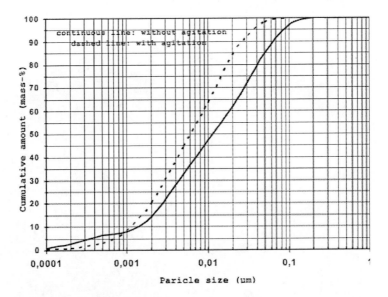

Figure 14. Laser particle size of tuffs.

According to Largent (1975), the potential pozzolanic activity of a material lies in silica content higher than 45%, a significant proportion of vitreous phase and a large specific surface.

These chemical, mineralogical and physical characteristics are found in the pumice tuffs studied, even if the vitreous phase is devitrified, because of the weathering they have undergone. The lime used in the stabilization of the tuffs is quicklime (CaO). Its characteristics are mentioned on table 8.

The manufacturing process of the phosphoric acid at the Chemical industries of Senegal (I.C.S) involves the formation of residue rich in gypsum which, after washing and drying, gives phosphogypsum. Figure 15 represents the phosphogypsum shown at SEM. The evaluation of the phosphogypsum stock shows a quantity available estimated at 166 million m^3. The chemical analysis of phosphogypsum is consigned in table 9. The phosphogypsum has a granulometry between 0 μm and 100 μm, with 30% of grains lower than 40 μm.

Table 8. Physicochemical characteristics of the lime

specific Surface Blaine (cm2/g)	Volumic masse (kg/m3)	Specific weight (Tonne/m3)	Hydraulicity Indices	Air hardening (%)
8.000 à 20.000	490 à 700	2.2 à 2.5	0 à 0,1	100

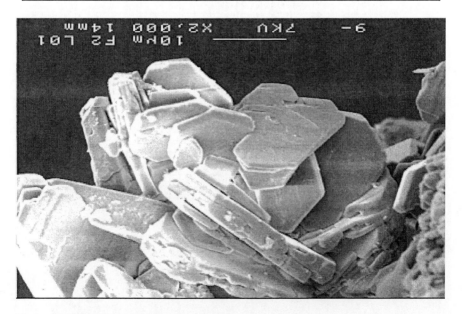

Figure 15. Senegalese phosphorgypsum at SEM.

Table 9. Average chemical composition of phosphorgypsum

Major elements	SO_3	CaO	SiO_2	P_2O_5	PF
Percentage	42.3	30.3	4.2	0.5	21.1

Table 10. Composition of the formulas

N° of the formulas	1	2	3	4
Tuffs (%)	100	96	96	94
CaO (%)		4	4	4
Phosphogypsum (%)				2

For the tests of manufacture (Table 10), four formulations containing tuffs and lime were studied. The phosphogypsum which is an activator is used in formula 4. The compositions of these formulas are described in table 4. According to Lea, (1974), the addition from 1 to 3% of gypsum to cement pozzolana lime can have sometimes a beneficial consequence on the development of the mechanical resistances but this effect is random and remains unforeseeable. To estimate the possibilities of application in road geotechnics, we used Proctor tests and CBR (Californian Bearing Ratio).

Table 11. Characteristic of the compressed earth bricks in Senegal (1: carrying structure, 2: noncarrying structure)

Category	Compressive strength dries (MPa)	Compressive strength wet (MPa)	Absorption of water (%)
1	≥ 2	≥ 1	
2	≥ 4	≥ 2	15

The pastes of the formulas 2, 3 and 4, conditioned in hermetically closed plastic bags, underwent maturation during 48 hours before there were compacted. Table 11 gives the standards of resistance for compressed earth bricks (BTC) in Senegal. The purposes of the tests on small bricks are to determine the ideal formulation. Three parameters must be optimized: content lime, water content and temperature. However, having noted in the preliminary tests an important swelling of the test-tubes due undoubtedly to the presence of swelling clays such as the illite (revealed by the X-ray diffraction of the tuffs), all the test-tubes are wrapped in aluminium paper in order to isolate them from ambient moisture. The cylindrical test-tubes have a diameter of 2 cm and are 4 cm height (figure 16). The optimization of mixing water consisted in varying the water content between 30 and 60% of weight of the solid matters (tuffs and lime). If water content is lower than 30%, the mixture is too thin (impossible to work). If water content is superior to 60%, the mix is too fluid.

Figure 16. Mould used to make cylinder bricks.

The optimization of the lime content consists in varying the percentage of lime between 1 and 20% of lime of weight of the solid matters. The temperature activates the pozzolanic reaction (Shi, 2001). The optimization of the temperature of maturation consists in varying the temperature of curing between 20°C and 120°C. For the tests on bricks, the principle of the formulations consists in using the minimum of lime and phosphogypsum while having bricks with physical and mechanical characteristics in conformity with the Senegalese standards. The formulas tested are consigned in table 12.

Table 12. Formulas used in the study (wt %)

Formulas	40	41	42	43	44	70	72	74	76	78
lime (%)	4	4	4	4	4	7	7	7	7	7
Phosphogypsum (%)	0	1	2	3	4	0	2	4	6	8
Tuffs (%)	96	95	94	93	92	93	91	89	87	85

The mixture is initially done with dry materials. Mixing must be quite neat. The mixture is controlled by appreciating the unit of the colour: no trail of lime nonbuilt-in the mixture must appear. The quantity of added water is fixed at 1500 grams, that is to say 30%. The whole is then well mixed until obtaining a very homogeneous paste. However, if the water content is too high (higher than 30%), the pressure of the compaction is deadened by the water which cannot be driven out among the grains. Thus, the press will not be able to allow the fabrication of bricks. The test-tubes are made using a Terstaram press provided with a mould of 22 side cm exerting a pressure rating equalizes with 21 kg/cm^2. This mould was then divided into two equal parts. What makes it possible to obtain two test-tubes of dimensions: 22 x 11 x 5.5 cm.

Results of the Tests

For the tests on small samples test bars, the graphics of figure 17 shows an asymmetrical bell-shaped curve with steep slope with an optimum of 50% for the water content. It is interesting to note that the optimum water content corresponding to the maximum density of the test-tubes is different; it is 40%.

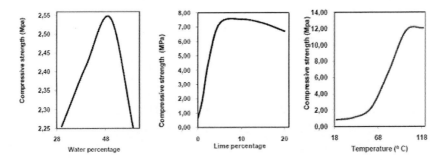

Figure 17. Optimization of the parameters of maturation Lime.

The optimum of lime (7%) is higher than the value generally found for clays (Perret, 1977) which is approximately 4% with the quicklime (CaO) and 5% with slaked lime (Ca(OH)$_2$). This difference is in connection with the extreme smoothness of the tuffs and their reactivity.

Here, it is important to note the pozzolanic character of the tuffs which, under the conditions of test, have a compressive strength of 5.3 MPa after three days of conservation only. The pozzolanic reaction starts around 57°C. But, it becomes truly effective only around 70 °C. Lastly, with 100 °C, a stage begins which seems to be maintained beyond 120 °C. Thus, it is useless to preserve the test-tubes beyond 100 °C. The study of the x-rays diffraction of (figure 18) shows in these weathered and devitrified pumice tuffs the following mineral paragenesis: amorphous quartz-hematite-kaolinite-illite-silica.

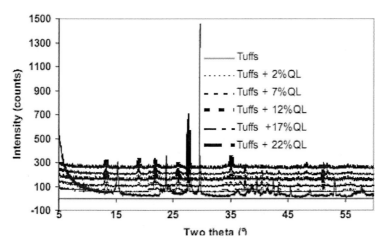

Figure 18. X ray diffraction of tuffs and tuffs treated.

The amorphous or vitreous phase is translated in x-rays by domes of diffusion Pichon (1992). Thus, according to the importance of the domes, the proportion of devitrified vitreous phase seems important. The X-ray diffraction of the tuffs and the tuffs treated with 2; 7, 12, 17 and 22% of quicklime (CaO) reveals a clear evolution of the mineral phases. The argillaceous minerals are definitely visible with the Scanning electron microscope. After treatment with lime, the peaks characteristic of kaolinite grow blurred with the addition of 2% of lime, proof that the pozzolanic reaction is quite effective. The argillaceous minerals and quartz are definitely less visible (reduction moreover 50% of the intensity) and all the peaks are shifted of approximately 1°. Lastly, the peaks characteristic of hematite almost disappeared. Compared to the lime addition, 7% corresponds to the most effective pozzolanic reaction. What confirms the results on small test tubs. At SEM, foams of de hydrated calcium silicate (HCS) are well visible (figure 20). The parameters of maturation of the test-tubes having been well controlled, long term tests are carried out in order to follow the evolution of the resistance of the test-tubes with time. In order to minimize the quantity of lime for economic reasons, we worked for the long term tests with lime 2.5%. The test-tubes are crushed after 2, 7, 14, 28 and 90 days of conservation. Figure 19 confirms the slow character of the expression of the pozzolanic effect. However, the performances obtained, higher than 4

MPa, shows that it is possible to work with very low lime contents. The calcined tuffs at 800 °C are less reactive than the natural tuffs.

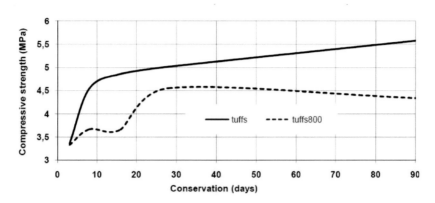

Figure 19. Mechanical test long term curing of bricks.

Figure 20. a) SEM of Mako tuffs: halloysite en choux-fleur, halloysite tubs, kaolinite book sheets and quartz grains; b) Felting of HCS; filamentous texture.

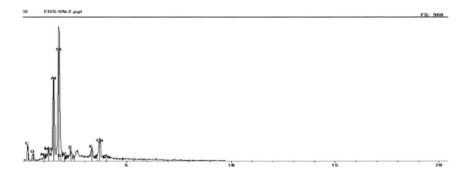

Figure 21. EDS of tuffs treated with 5% by weight of lime.

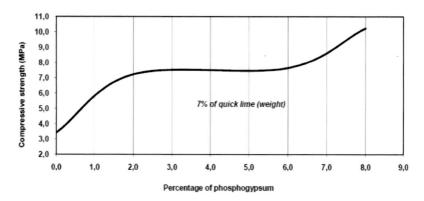

Figure 22. Variation of resistance according to the content phosphogypsum to lime 7% with 60°C.

Figure 21 represent EDS of tuffs treated with 5% by weight of lime. The principal elements are silica, alumina and calcium. Figure 22 corresponds to the compressive strengths of the test-tubes made with a fixed content of lime equal to 7%, after 7 days of cure in oven at 60 °C. The compressive strengths obtained are high. According to the figure, the compressive strength increases gradually with addition of phosphogypsum. A stage seems to take shape between 2 and 6% of phosphogypsum. Then, the curve presents an ascending slope.

Table 13. Values of CBR obtained with the formulas

	Requirements layers		N° of the formulas			
	Fond	base	1	2	3	4
CBR (%): 3 days in the air and 4 days of imbibitions	6,0 - 80	≥ 160	8	30		
CBR (%): 7 days at 60 °C	80 - 120				140	195
CBR (%): 7 days at 60 °C and 4 days of imbibitions	6,0 - 80	≥ 160			158	210
Compressive strength: 3 days in the air and 4 days of imbibitions (bars)	2.5 - 5,0	≥ 5,0				
Compressive strength 7 days in the air (bars)	5,0 - 10	18 - 30				

Table 13 shows the results of the compaction tests. The value of the CBR of the tuffs alone is low (8%) and allows its use in road geotechnics only in sub-base. With a treatment of 4% of lime, the bearing capacity of the tuffs is clearly improved, the CBR is then equal to 30% but the use of such a material is limited to the sub-bases of road.

With lime, when the moulds are heated at oven at 60 °C, the value of the CBR is multiplied by more than 10 (140%). The phosphogypsum addition, which is an activator, creates a gain of 40% on the CBR (195%). The tuffs treated with lime and activated with phosphogypsum are usable in road base. For formula 4, we carried out a punching after 4 days additional of imbibitions. We note an improvement of the CBR of 8%. Formula 4 activated with 2% of phosphogypsum presents a very high CBR (210%).

Interpretation of Results

The silica and alumina released during the dissolution of argillaceous minerals under the effect of the quicklime react with calcium in solution to form by precipitation hydrated calcium silicates (HCS) and hydrated calcium aluminates (HCA). The role of the HCA in the binders pozzolana-lime is always additional because cohesion resulting from the pozzolanic activity is mainly due to the HCS not very basic and rich of water. Indeed, the former experiments (Desgache and Piquel, 2003) highlighted a more important rule of Si/Al ratio in the products of the pozzolanic activity than the basic materials. With the temperature lower than 40 °C, bricks kept in the air, the compressive strengths obtained are weak. This is explained why the pozzolanic reaction proceeded partially. For reaction to be completed, it is necessary to have a temperature of cure higher or equal to 40°C. The technique of thermal activation consists in increasing the temperature of cure by using an oven or a cover in plastic polyane. The chemical technique of activation consists in adding phosphogypsum to the mixture. Phosphogypsum adjunction to the mixture presents several effects, in particular the increase in the mechanical resistance. This confirms the work of (Lea, 1974). The phosphogypsum addition causes a light increase in density. The phosphogypsum addition does not seem to improve the sensitivity to water of the test-tubes. The ratio of resistances dry and wet is constant and equal to 2.2 as well as for the formula with phosphogypsum and without. The calcined tuffs at 800 °C are definitely less reactive than the natural tuffs, whatever the term. The transformation of the kaolin into metakaolin (more reactive) starting from 550 °C lets predict the reverse. Moreover, former work (Le Borgne and Meyer, 2005) showed that clays could represent until nearly 30% of the normative composition of the tuffs and the calcinations of vitreous materials generally increases their fineness. The explanation of the phenomenon could lie in the fact that the iron hydroxides are transformed into less reactive oxides. Lastly, the phosphogypsum addition to the mixture causes neither the creation of new minerals, nor a change of texture of bricks. However, observation of the photographs carried out under the Scannning Electron Microscope (MEB) shows the presence of calcium silicate hydrate foams (HSC) quite visible. One can deduce from it that the crystals observed with the naked eye consist of calcium silicate hydrates and hydroxide of calcium. The ettringite and the monosulfo-aluminate could be present when the calcium sulphate (phosphogypsum) is added to the reagents (Desgache and Piquel, 2003), but these two minerals have not been observed on our stereotypes. The passage of a vesicular texture to a filamentous texture results from the pouzzolanique reaction between silica, alumine and the quicklime. For road compaction tests, with sampes treated with lime, when the moulds are conditioned in the oven at 60 °C, the value of the CBR is multiplied by more than 10 (140%) because of thermal activation of the pouzzolanique reaction. The phosphogypsum addition which is an activator creates a profit of 40% on the CBR (195%). The presence of the activator (phosphogypsum) accelerates the pozzolanic reaction and makes it possible to reach the required values quickly. It is important to note that these two formulas containing lime after one week of heat treatment resist to the action of water perfectly. At this time, the pozzolanic reaction definitively took the step on swelling. Water becomes thus a medium favourable to the kinetics of the pozzolanic reaction.

Conclusion

The pumice tuffs of eastern Senegal suit well to a treatment by lime. The treatment by cement is not economically possible. However, for the tuffs treated with lime, it is necessary to avoid their contact with water the first days of cure to avoid the swelling of clays. In Eastern Senegal, the temperature borders and exceeds 50°C as soon as the month of march. What allows a natural thermal activation of the pozzolanic reaction. However, as this one is slow, the use of phosphogypsum as chemical activator makes it possible to have quickly a material in conformity with the Senegalese standards on compressed earth bricks for construction and road geotechnics. This in addition makes it possible to avoid stockpiling of this phosphogypsum and its fatal consequences on the environment. Also, the spaces occupied before by these heaps can then be reallocated with other uses, in particular agricultural.

References

Prayon Rupel Technologies (1996) : Section conversion et qualité des sous-produits de sulfate de calcium ex-phosphate de Taïba - *Test CPP (Central Prayon Process).*

Coatanlem, (2004) : Pouzzolanicité des tufs volcaniques acides du Sénégal oriental, XXIIe RUGC 2004, Ville and Génie civil, 3 and 4 juin 2004, Marne-la-Vallée.

N'Diaye, Matar et al. (2003): « Pouzolanic activity of acid and intermediary volcanic tuffs of Mako areas (Senegal) », International symposium on industrial minerals and building stones, September 15-18.

Pichon H. (1992) : Le système « Pouzzolane naturelle chaux » á 38 et 100 °C, relation entre la composition chimique, les phases néoformées et les conséquences physico-chimiques, Thèse de doctorat, Université Joseph-Fourrier, Grenoble.

Largent (1975) : Estimation de l'activité pouzzolanique. Recherche d'un essai. Bull. Liaison Labo. P. et Ch., 93, janvier-février, p. 61-65.

Lea, F. M. (1974): The chemistry of cement and concrete, 3rd Edition, Edward Arnold, New York.

Caijun Shi, (2001): An overview on the activation of reactivity of naturals, *Can. J. Eng*, 28, 778-786.

Perret P, (1977) : Contribution à l'étude de la stabilisation des sols fins par la chaux : étude globale du phénomène et applications. Thèse de Doct. Ing, INSA de Rennes, 166 p.

Fournier M. et Geoffray J. M., (1978) : Le liant pouzzolanes-chaux. Bulletin de liaison du laboratoire des Ponts et Chaussés, n° 93, réf. 2145, LCPC.

Le Borgne T., Meyer S. (2005) : Pouzzolanicite de tufs volcaniques ; fabrication de blocs de construction pour l'habitat traditionnel ; Master Pro aménagement ; Université Blaise-Pascal, Clermont-Ferrand.

Desgache G., Piquel J. (2003) : Etude de cas : Mise en évidence de l'activité pouzzolanique de deux matériaux volcaniques du massif central, DESS de géologie de l'aménagement, Université Blaise-Pascal de Clermont-Ferrand, UFR Sciences exactes et naturelles, 51 p.

B. Alkali Activated Clay Material Bricks

Abstract

The paper discusses the addition of sodium hydroxide to readily available natural materials in Senegal, in order to produce stronger and more durable bricks. The case study involves the natural clays (Tchieky, allou Kagne, Niemenike), highly weathered rocks (laterite) volcanic tuff (Bafoundou) and mining waste (aluminum phosphate, calcium phosphate), which were mixed with varying amounts of NaOH and heated at temperatures of 40, 80 and 120°C for periods of up to 90 days. The caustic solution could be sodium hydroxide mixed with sodium silicate and water/seawater. In general the maximum strength was reached after 28 days for samples mixed with 8 M NaOH at 40°C/60% relative humidity. The bricks produced were shown to be durable and relatively inexpensive. The process does not generate pollutants and can use by-products, assisting with environmental problems.

Keywords: Sun-dried bricks. Alkali activation. Durability. Strength. Zeolites. Geopolymer, Senegal

Introduction

In 1970s Davidovits (Davidovits, 1989) pioneered the discovery and establishing the research in geopolymer binders. Geopolymers are a novel class of materials that are formed by the polymerization of silicon and aluminum species. The principal binding phase in geopolymers is an amorphous aluminosilicate gel that consists of a three dimensional framework of SiO_4 and AlO_4 tetrahedra linked by corner-shared O atoms (Davidovits, 1989; Provis et al.,2005; Slavik et al.,2005). The negatively-charged tetrahedral Al sites in the network are charge-balanced by alkali metal cations such as Na^+ and/or K^+ (Davidovits, 1989). However, sodalite and hydroxysodalite, which are members of zeolite group, have been detected as reaction products in some metakaolin and fly ash geoplolymer systems (Slavik et al., 2005; Palomo et al., 1999; Slavik et al. 2005).

Geopolymer binders possess many advanced properties such as fast setting and hardening, excellent bond strength (Davidovits, 1989), long-term durability, better fire and acid resistance (Phair and Deventer, 2001). Due to such superior property, geopolymers have the potential to be used in several industrial applications (Xu Hua and Van Deventer, 2000). The most important advantage of geopolymer binders is its low manufacturing energy consumption and low CO_2 emission (Xu Hua and Van Deventer, 2000; Davidovits, 2005; Davidovits et al. 1990), which make it to be a "Green Material" (Davidovits, 1999; Davidovits et al. 1999; De silva et al., 2007). The original raw material used by Davidovits is pure metakaolinite, activated by alkali hydroxide and/or alkali silicate (Davidovits, 1989; Davidovits, 2005; Davidovits et al. 1990). Many researchers (Xu Hua and Van Deventer, 2000; Xu and Van Deventer, 2002; Phair and Van Deventer, 2002; Komnitsas et al., 2007; Panias et al., 2007) have been demonstrated that many other aluminosilicate materials could be used as raw materials for geopolymers, such as fly ash, furnace slag, silica fume and kaoline and some natural minerals. (Xu Hua and Van Deventer, 2000) investigated geopolymerization of sixteen natural aluminosilicate minerals with the addition of kaolinite. It was found that a wide range of natural alumino-silicate minerals provided potential sources for synthesis of geopolymers.

It is not a resource and energy saving process to prepare geopolymer from pure metakaolinite as raw material (Yang and Zeng, 2005). Although there are a shortage of pure kaolinitic raw materials, but there are abundance in low value clays raw materials (low koalinitic clays) all over the word. Thus exploiting these raw materials in production of binder is an advantage. However, even if these binders cannot fulfill the requirements of cement specifications, due to low active component, their activity can be exploited in many other construction materials like bricks and board composites. The later may be an economical alternative for developing countries because the technological development required is very small. However, the chemical composition of low kaolinitic raw materials is very complicated. Impurities existing with kaolin in low value clays may complicate the study of the geopolymerization process (Yip et al., 2008; Rahier et al., 2007; Hos et al., 2002; Brew and MacKenzie, 2007). Thus, investigating the mechanism of geopolymerization of these low value clays is essential.

Freidin (2007) and Yang et al. (2008) used geopolymerisation of waste by-products to produce cement less pressed blocks. Bassir Diop et al (2008) developed a low temperature bricks based on geopolymerisation of tuff using inexpensive processing. Billong et al. (2009) used thermally-treated lateritic soil partially activated with sodium hydroxide (NaOH) in the production of compressed blocks.

Senegalese Clay Texture Material Used

Available Materials

Varieties of clay (Tchieky, Allou Kagne, Niemenike), highly weathered rocks (laterite,) volcanic tuff (Bafoundou), and mining waste (aluminum phosphate, calcium phosphate) are materials available in Senegal for making brick.

The Tchieky deposits in western Senegal (Figure 23). They belong to the Cap Rouge Series from the top of the Ndiass horst of Cretaceous age. Petrographically, the clay is homogeneous and grey in color.

Allou Kagne attapulgite deposit in western Senegal (Figure 23) belongs to the series of Cap Rouge which constitutes the stratigraphical level of the top of the Horst of Ndiass dating from the end of the cretaceous. Petrographically, this clay is homogeneous and gray of colour.

Niemenike is located in southeastern Senegal near the town of Kedougou (Figure 23). The clay of Niemenike belong to the Birrimian-period (2 billion years) and are alteration products of granitic rocks. Petrographically, this clay is homogeneous and yellow of colour. We compare the performance of clay from the Niemenike deposit and clay pre-treated at 700 °C.

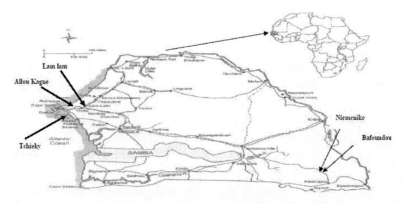

Figure 23. Map of the African continent showing the location of Senegal. The enlargement of Senegal shows the geographical location of Tchieky, the Allou Kagne attapulgite, Niemenike, Lam lam and Taiba samples used in the research.

Procedure

The process of making zeolites from metakaolin is straight forward (Palomo et al, 1999a); (Palomo et al , 1999b). Metakaolin ($Al_2Si_2O_7$) and Class F fly ash have the correct SiO_2 to Al_2O_3 ratio needed to synthesize zeolite A (Na_{12} [$Al_{12}Si_{12}O_{48}$] $27H_2O$). If one wishes to make zeolite X or Y one must add additional silica either as sodium silicate or finely divided amorphous quartz. Metakaolin is normally mixed with sodium hydroxide and allowed to react in large boiling-water heated vessels. Water-rich slurries are stirred until zeolites form and the reaction goes to completion. Monolithic samples can also be produced, but one must limit the amount of liquid used to a bare minimum. In this case, stoichiometric amounts of 4–8 M NaOH are mixed with the dry ingredients to form a thick putty-like paste that is then molded and cured at elevated temperatures. Monoliths made in this way are generally very strong and highly insoluble.

These particular clay texture material have been characterized using a variety of techniques including chemical and physical analyses. DTA/TGA, laser particle size determination, X-ray diffraction, SEM microstructure were performed with energy dispersive spectra (EDS).

The work reported here is an outgrowth of our previous work, but what is new is the fact that the above process also works using the clay texture material without any thermal pre-treatment of the clay. Although there are scattered references to rather slow reactions of kaolinite with caustic (Bao and al, 2005) (Berg et al, 1965) that tend to reinforce current thinking that the reactivity of clays must be increased by heating them to 500-700 °C (i.e. dehydroxylating them) to increase their reactivities and make their use viable in a real time frame (Berg et al, 1968) the outcome of the work with the clay texture material was unexpected and as such not previously documented.

All samples are 2.5 cm in diameter and 5.0 cm long. Note that additional samples were allowed to age for extra time – 90 days total – in order to obtain better long term. Because the clay texture material samples are partially agglomerated, the starting material was ground to less than 250 μm. The clay texture material was mixed with different alkali concentrations (4, 8 and 12 molar NaOH) to form thick pastes. See Table 14. After vigorous hand mixing, the bricks were compressed in a die (2.5 mm in diameter) with a hand-operated hydraulic press (Figure 24). Pressure was applied until water began to be ''squeezed'' out of the sample. Pressures were typically in the 10 MPa range. The cylinders were trimmed to 5.0 mm in length and then cured at three different temperatures (40° C, 80° C and 120 °C). The 40 °C samples were cured in a ''walk in'' chamber that was maintained at 60% relative humidity. The 80 °C samples were cured in an oven in a desiccator that was only partially sealed so that the samples gradually lost what extra water they had over time. The 120 °C samples were cured in sealed Parr type vessels fitted with Teflon liners at 120 °C for varying periods of time. All cylinders were allowed to sit overnight at room temperature before they were cured. This so called "soaking" is typically used to allow time for dissolution and zeolite precursors to form (Palomo et al, 1999a); (Palomo et al, 1999b). After curing as a function of time, the mechanical behavior of the cylinders was tested. For the samples cured at 120 °C, compressive strength values were measured after 6, 12 and 24hours, because the hydration kinetics at 120 °C was significantly greater than at either 40° or 80 °C. For samples cured at 40°C/60RH and 80 °C, cylinders were tested after 4, 14, 28 and 90 days. In order to test durability, pieces of the three samples cured at 120 °C for 12 h (#s 2, 15, 28) were ground to size (less than 150 and more than 75 μm) and dried at 105 °C.

Figure 24. Hand-operated hydraulic press used to compact bricks.

Table 14. Formulations studied

Sample	Clay material materialTuff (wt%)	NaOH (wt%)	Cure temperature (°C)	Cure time (days)
1	80	4 molar	120	0.25
2		(20)		0.5
3				1
4	80	4 molar	40	7
5		(20)		14
6				28
7				60
8	80	4 molar	80	7
9		(20)		14
10				28
11				60
12	80	8 molar	120	0.25
13		(20)		0.5
14				1
15	80	8 molar	40	6
16		(20)		12
17				24
18				60
19	80	8 molar	80	7
20		(20)		14
21				28
22				60
23	80	12 molar	120	0.25
24		(20)		0.5
25				1
26	80	12 molar	40	7
27		(20)		14
28				28
29				60
30	80	12 molar	80	7
31		(20)		14
32				28
33				60

One gram of each of the sized materials was placed in 10 mL deionized water and held at 90 °C for 1 and 7 days in a sealed Teflon container. The test is a modified product consistency test (PCT) designed to test glass leaching (ASTM C1285). These samples were chosen because it was assumed that reaction had proceeded to the greatest degree at 120 °C and leaching of these samples would better reflect what would happen to all brick that had been cured for a longer time once it was used to build a house (many years).

Results

In Table 15 are represented the chemical composition of natural clay studied in weight percent. All clay are rich in silica, the major oxides are SiO_2, CaO, Al_2O_3 and Fe_2O_3. Allou Kagne clay is rich in MgO. The representation of these clays on a ternary diagram system SiO_2-Al_2O_3 –Na_2O (Figure 25) shows that they are close to the silica pole and need variable amount of caustic to be close to the zeolite domain. The attapulgite clay is extremely fine with an average grain size equal to 13.73 μm. Particle size was determined using laser granulometry with and without ultrasonic treatment (Figure 25). Although referred to as "clay", the Tchieky deposits have 64% silt fraction; the average grain size being 4.79 μm.

**Table 15. Chemical composition of Tchieky, Allou Kagne
and Niemenike clay (wt %)**

Oxides	SiO_2	CaO	Al_2O_3	Fe_2O_3	MgO	K_2O	Na_2O	TiO_2	MnO	P_2O_5	LOI	Reactive silica
Tchieky	62.1	0.14	13.6	8.39	0.73	1.22	0.13	1.05	0.03	0.16	12.33	-
Alou Kagne	44.9	9.65	4.2	2.43	10.3	0.30	0.06	0.22	0.01	0.54	27.4	-
Niemenike	69.7	13.30	10.1	0.10	0.10	1.30	0.60	0.80	0.10	0.03	4	8.9

The Niemenike clay is the most quartz- rich because it form directly from granitic rocks and there was any transportation of weathered products. Niemenike is also rich in CaO and Al_2O_3 connected with clay mineral.

The deformation curve of the cylinders shows that the rupture is progressive. Breaking of the 2.5 by 5.0 mm cylinders during compression was good; all breaks exhibited a typical double pyramidal shape (Figure 26). The Stress-Strain curves represented on figure 26 corresponds to Tchieky clay treated with 4, 8 and 12molars NaOH solution and kept in chamber at 40°C/60% relative humidity and at oven at 80°C during 60 days. Stress–Strain curves of alkali activated Attapulgite clay are represented on figure 10 and those of Niemenike on figure 14. Compressive strength curves of alkali activated clay samples from Tchieky, Allou Kagne and Niemenike are represented on figures 12, 13 and 14 respectively.

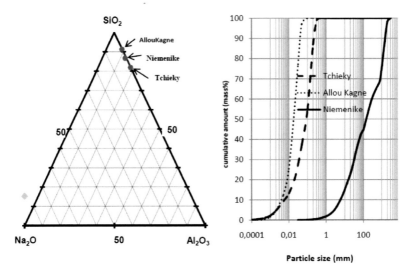

Figure 25. Ternary composition diagram for the system SO_2-Al_2O_3-CaO (view (right)) and Particle size analysis (view (left))

Figure 26. Typical breaks of samples with double pyramidal shape.

Typical stress–strain curves for the Tchieky clay samples made with 4, 8 and 12 M NaOH solutions and cured at 40°C/60%RH and 80°C for 60 days are given in figure 28. A summary of the strength data is given in figure 27, which indicates that the highest strengths are recorded for all the samples at 28 days whether 4, 8, or 12 M NaOH was used. With 4 and 8 M NaOH, the samples cured at 40°C/60% RH gave the maximum strengths while for the 12 M mix, the highest strength was for samples heated at 80°C/low RH.

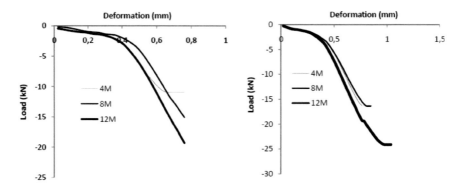

Figure 27. Stress strain curve of alkali activated Tchieky clay kept at Chamber (view right) and 80°C (view left) during 60 days.

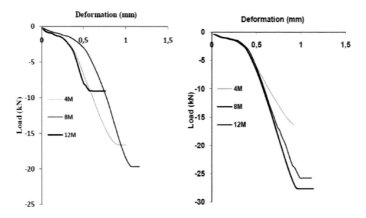

Figure 28. Stress strain curve of alkali activated Attapulgite clay kept at chamber 40°C/60%RH during 60 days and at 120°C (6, 8 and 12 hours).

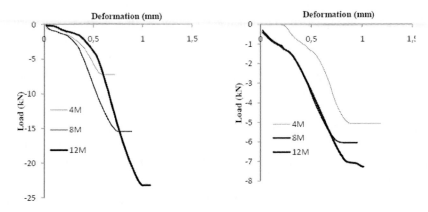

Figure 29. Stress strain curves for the Niemenike clay samples cured at 40 °C/60%RH (View right) for 60 days and at 120 °C/0%RH (View left) for 12 hours. Each sample was made with 4, 8 and 12M NaOH solution as a thick paste and then cured.

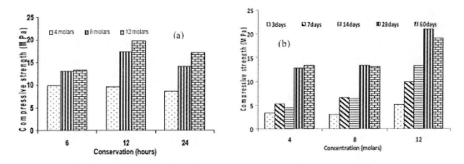

Figure 30. Compressive strength of Tchieky alkali activated clay brick 120°C (View a) and 40°C/60%RH (View b).

Samples fabricated from Tchieky clay and NaOH solutions attain compressive strengths ranging from 7.5 to more than 21 MPa (Figure 30), similar to those obtained on samples made with meta-kaolin. The maximum strength obtained with the three type of conservation are comparable: 19,7 MPa for the chamber, 21,1MPa for the oven at 80°C and 20,8 MPa for samples kept at 120°C. The examination of the results shows that the most advantageous process is to activate the clay with 8 molar NaOH solutions and conserve in a chamber (40°C/60°RH) for one month.

The increase in strength is dependent on temperature and length of curing, alkali concentration, fineness, crystallinity and composition of the raw materials. For samples cured at 120°C, the best mechanical performance is obtained after 12 h. The strength gain at any given time is directly proportional to the concentration of NaOH used. For reactions carried out at lower temperatures, samples that were cured in the 40°C/ 60% RH chamber attained higher strengths than samples cured at 80°C for the first month for 4 and 8 M concentration. It is believed that the presence of moisture in the 40°C chamber allowed hydration reactions to continue for longer than the equivalent sample cured at 80°C in a ''semi-dry'' atmosphere.

Samples fabricated from Allou Kagne attapulgite clay and NaOH solutions attain compressive strengths that range from 4.5 to more than 27 MPa (Figure 31). These strengths are in the range of similar samples made with metakaolin. Strength development is dependent on temperature and length of curing, alkali concentration, fineness, crystallinity and composition of the raw materials. For samples cured at 120 °C, 24 h temperature seems to impart the best mechanical performance to the samples regardless of the concentration of NaOH solution used to make the sample. It is also notable that the strength gained at any given time is directly proportional to the concentration of NaOH used to make the sample. For reactions carried out at lower temperatures, samples that were cured in the 40°C/60% RH chamber attained higher strengths for the first month for 4 and 8molars concentration.

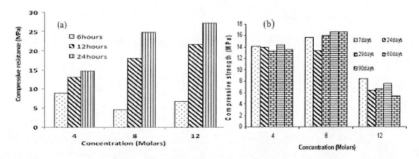

Figure 31. Compressive strength of alkali activated Allou kagne clay brick samples cured at 120 °C (view a) and 40 °C (view b).

For samples kept on chamber at 40°C/60% RH, the maximum strength is obtained after three months with 8 molars NaOH concentration (16,7MPa). For the chamber, the most interesting process is to activate the clay with 8 molars NaOH concentration for 7 days (15,7MPa). If quick process is needed, the most advantageous method is to activate the clay with 8 molars NaOH concentration for 24hours (24,7MPa).

For attapulgite clay, the best compromise seems to activate the clay with 8 molars NaOH concentration for 7 days in the chamber (15,7MPa).

Samples fabricated from Niemenike clay and NaOH solutions attain compressive strengths that range from 6 to more than 16.9 MPa (Figure 32). These strengths are in the range of similar samples made with metakaolin.

For niemenike clay, summary of the strength data is given in figure 32 for both clays. The curing time of the 40 °C samples was extended to 60 days. Interestingly, the composition of the NaOH solution did affect the outcome of the 60 days of runs.

Figure 32 gives the summary of compressive strength of samples cured at 40 °C and 120 °C.

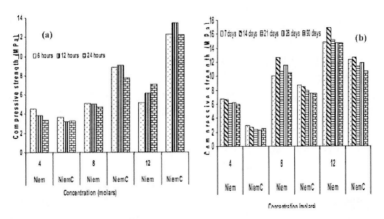

Figure 32. Compressive strength of alkali activated Niemenike clay brick samples cured at 120°C (view (a)) and 40 °C (view (b)).

In long-term tests for bricks kept at 40C/60% RH, strength did not increase with time for both (natural and calcined) clays activated with sodium hydroxide. The maximum strength was obtained after 14 days for all concentrations. For all concentrations and periods, strength obtained with natural clay was greater.

For all bricks, strength increased with concentration. The maximum strength obtained with 4M concentrations was 6.9 MPa for natural clay and 2.9 MPa for calcined clay after 1 week.

For 8M concentration, the maximum strength obtained with the natural clay was 12 MPa after two weeks and 8.6 MPa with the calcined clay after the same period.

For 12M concentration, the maximum strength obtained with the natural clay was 16.9 MPa after two weeks and 12.6 MPa with the calcined clay for the same period.

In short-term tests, for bricks kept at 120 °C/60% RH, the situation changed; except for 4M concentration, the calcined clay gave the best mechanical performance. For all bricks, strength increased with concentration.

For 4M concentration, the period of 6 hours gave the best performance for the natural clay as well as the calcined clay.

For 8M concentration, the strengths obtained after 6 hours of curing were close to those obtained after 12 hours.

For 12M concentration, the maximum strength is obtained after 24 hours of curing for the natural clay (6.1 MPa) and 12 hours (13.4 MPa) for the calcined one.

For long-term tests, bricks were kept at 40C/60% RH. There was no need to calcine the clay because the natural clay showed the greatest strength regardless of the concentration and the curing time. Also, there was no need to cure bricks for more than 2 weeks. Breaking of the 2.5 by 5.0 mm cylinders during compression was good; all breaks exhibited a typical double pyramidal shape. For Tchieky clay, Figure 37 (top a and b) represents the X-ray diffraction patterns for the original material and for the mix with 12 M NaOH after curing for 7 days at 40, 80 and 120°C (top b). It can be seen that there is little difference between the patterns. A general broadening of the kaolinite peaks indicates that some disordered phases are forming which are clay-like but lack the crystallinity of kaolinite. Traces of lizardite are also present. The poorly crystallized phases may be the cementing phases that give the sample its strength. The plots for the Allou Kagne attapulgite clay are represented later in figure 37 with accompanying reaction products of clay samples activated with 12 molar NaOH solutions (middle b) cured at 80 °C for 7 days. The sample contains: quartz, kaolinite, illite, hematite and muscovite (middle a). There is a general broadening of the palygorskite peaks which indicates that some disordered phases that are clay-like are forming, but lack the crystallinity of palygorskite. The poorly crystallized phases may be the cementing phases that give the sample its strength. Figure 37 represents the X-ray diffraction patterns for the starting material and the patterns for the 40 and 120 °C samples cured for 7 days made with 12 Molar NaOH solutions for the Niemenike clay. X-ray diffraction shows this following paragenesis: quartz, kaolinite, nacrite, halloysite, hematite and muscovite. The peak at 24° is thought to be associated with sodalite ($Na_4Al_3Si_3O_{12}Cl$). Muscovite ($KAl_2Si_3AlO_{10}(OH)_2$) is also present and the 12M concentration seems to contain some halloysite $(Al_2Si_2O_5(OH)_4)$ and possibly bucchulite $(Ca_2Al_2SiO_6(OH)_2)$ Halloysite tubs have been seen on SEM photos on Bafoundou tuffs (Diop and Grutzeck, 2007). Bafoundou is located three kilometers flying distance at the south of Niemenike. Figure 33 represents the microstructure typical of alkali activated Tchieky clay brick. The sample was made with 12M NaOH and cured at 40°C/60 %RH for 60days.

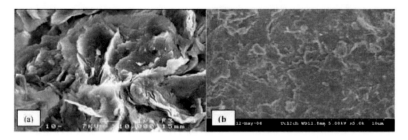

Figure 33. Views represent typical microstructures of Tchieky clay and the same clay treated with 12 M NaOH solution and cured for 90 days at 40 °C and 60 %RH.

The microstructure is more plate like than the equivalent sample made with 4 and 8 molar NaOH which are more granular in appearance. Figure 34 represents the microstructure typical of alkali-activated Allou kagne attapulgite clay brick. The sample was made with 12 M NaOH and cured at 40 °C for 60 days. Views (a) shows the phases present (b) are fiber of palygorskite. The microstructure is more plate-like than the equivalent samples made with 4 and 8 M NaOH, respectively. These latter samples are more granular in appearance that seems to reflect the greater degree of reaction of these samples caused by the higher alkalinity they contain.

Figure 34. Views represent typical microstructures of Allou Kagne attapulgite clay. Views (a) shows that the phases present are fiber of palygorskite. Views (b) represent brick samples made with 12 M NaOH solution and cured for 90 days at 40 °C and 60 %RH.

Figure 35 represents the microstructure typical of the alkali-activated
Niemenike clay bricks. Views (a) shows the phases present. Views (a) are
solid quartz (center) and mainly books of kaolinite. The sample was made with
12 M NaOH and cured at 40 °C for 60 days. The microstructure is more plate-
like than the equivalent samples made with 4 and 8 M NaOH, respectively.

Figure 35. Views represent typical microstructures of Niemenike clay. Views (a)
shows the phases present are solid quartz (center) and mainly books of kaolinite.
Views (b) represent brick samples made with 12 M NaOH solution and cured for 90
days at 40 °C and 60 % RH.

These latter samples are more granular in appearance, which seems to reflect
the greater degree of reaction of these samples caused by the higher alkalinity
they contain. Energy Dispersive Spectrum (EDS) on Niemenike (Figure 36)
clay which is a weathering of alkaline granitic shows spectrum corresponding
to a muscovite mineral. Sometimes spectrum corresponds to a quartz mineral
covered by fine clay or reflect a clay mineral: kaolinite ($Al_2Si_2O_5$ $(OH)_4$) or
halloysite ($Al_2Si_2O_5$ $(OH)_4$ $2H_2O$). Around five spectrums are on books of
kaolinite.

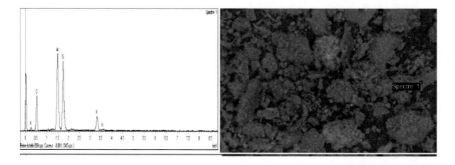

Figure 36. EDS of Niemenike clay.

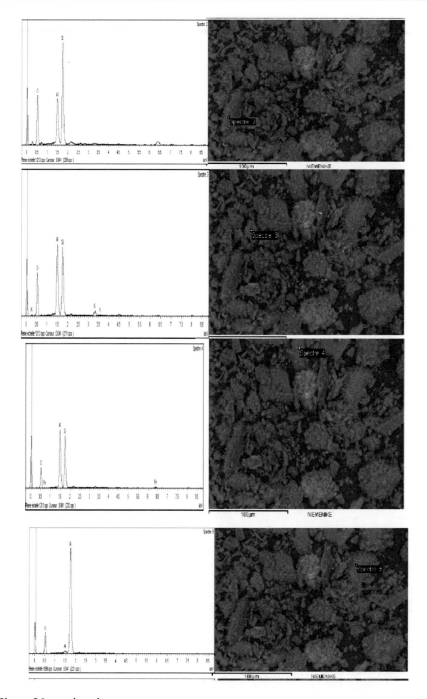

Figure 36. continued.

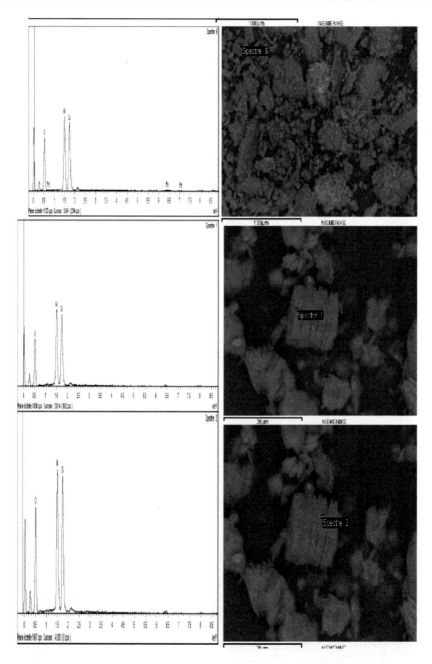

Figure 36. continued.

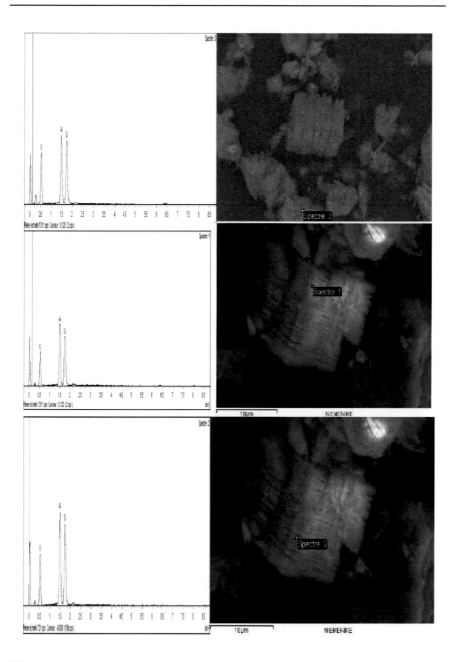

Figure 36. continued.

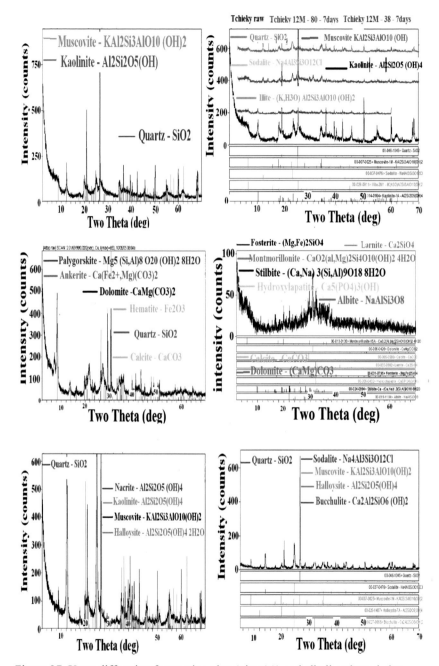

Figure 37. X ray diffraction for starting clay (view(a)) and alkali activated clay (view(b)) for Tchieky, allou Kagne and Niemenike.

Table 16. Results of leaching tests of bricks of Tchieky and Niemenike clay cured at 120°C

Concentration (molar)	Sites	Measurement at 1 day		Measurement at 7 days	
		Conductivity (mS/cm)	pH	Conductivity (mS/cm)	pH
4	Tchieky	1.40	10	1.10	10
	Niemenike	0.30	10	0.50	10
8	Tchieky	1.70	10	1.60	10.5
	Niemenike	0.10	10	0.30	10
12	Tchieky	2.70	11	2.50	11
	Niemenike	0.10	10	0.30	10

After 24 h, the leaching tests on the 120°C samples showed very low conductivities for all mixes regardless of the concentration of alkali used to make the brick. These values decreased with time for Tchieky clay, (Table 16), with 1 day conductivities greater than 7 days values for all three samples. But for Niemenike clay, the values of conductivities increased slightly with time. There is a kinetic process (possibly diffusion controlled) that limits the build-up of Na in solution. It is the sodium which accounts for the conductivity; Bao et al. (2005) suggest a standard solution of NaOH in water with a conductivity of 1 mS/cm contains 200 ppm NaOH. The 4 and 8 M samples have the lowest overall conductivities, whereas the 12 M samples are almost twice as high. This suggests that the amount of Na in the 4 and 8 M brick reacts nearly completely, whereas the 12 M sample may contain excess NaOH or soluble sodium silicate which washes out giving it a higher conductivity and pH. Nevertheless, all conductivity values are reasonably low, proving that reactions are occurring during curing and that zeolite-like mineral(s) are probably forming (Bao and Grutzeck 2006).

Senegalese Phosphate Waste

Introduction

The aluminum mining waste (aluminum phosphate), called feral is produced during the processing of a phosphate rich alumina (25wt% P_2O_5, 27wt%Al_2O_3, 9.53 wt% Fe_2O_3) which makes up the Lam-Lam deposit in western Senegal (Figure 23). Lam-Lam is one of the few aluminium phosphate deposits in the world. During processing, 30% of the phosphates are disposed of as waste (0/3mm in size). The waste is called feral due to its high iron and aluminum oxide content: 10 and 27 wt % respectively.

It is anticipated that approximately 15 million tons of feral will be produced in the coming years. This is a significant amount of waste material. Unfortunately Senegal, a developing country in West Africa, does not have the money or technology to recycle these wastes. Thus they languish in lakes and ponds where they have the potential to impact the environment in a negative fashion. In this note, we explore the possibility of using Feral to make bricks for social housing.

Phosphate mining waste called schlamm from Taïba deposit in western Senegal (Figure 23). Taïba is a calcium phosphate deposit. During processing, 30% of the phosphates are waste of clay size particle (0-40μm). In the short term, it's around 15 million tons of this schlamm which will be pile up with its negative consequence in the environment. Senegal a developing country in West Africa does not have the technology to recycle these wastes. In this note, we explore the possibility of using it to make bricks for social housing.

Results

Table 17 gives chemical composition of feral and schlamm samples in weight percent. Feral and schlamm mining phosphate waste have Al_2O_3 and P_2O_5 content P_2O_5 that enable the formation of AlPO4 zeolitic type.

Table 17. Chemical composition of feral and schlamm samples in weight percent

Samples	SiO_2	CaO	Al_2O_3	Fe_2O_3	MgO	K_2O	Na_2O	TiO_2	MnO	P_2O_5	BaO	SrO
Feral	12.5	6,7	27	9,5	0.1	0.1	1,3	2,0	0.1	24,9	0,1	0,4
Schlamm	22,7	30,5	8,7	3,8	1,1	0,3	0,1	0,4	0,1	22,2	0,02	0,1

DTA/TGA data for the attapulgite are presented in 39. These curves resemble those obtained when a sample of kaolin (kaolinite plus quartz) is first heated and then cooled. Kaolinite contains both adsorbed and interlayer water that come off gradually at 100-400° (zeolitic water) and sharply at 500 °C (structural), respectively. The adjacent sharp peak on both the up and down curves (DTA) is most likely due to the a–β quartz transition.

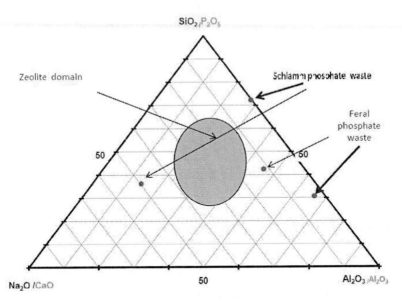

Figure 38. Ternary composition diagram for the system SiO_2-Al_2O_3-Na_2O and SiO_2-Al_2O_3-CaO.

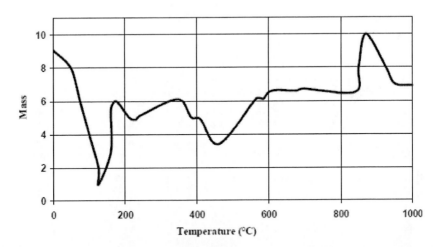

Figure 39. DTA/TGA data for the attapulgite.

Table 18 gives physical and mechanical properties of feral.

Table 18. Physical and mechanical properties of feral

Grain size	
% > 2mm	10
% < 80μm	24 (26 after CBR)
% < 2μm	13
Cu	600
Cc	32
Sand equivalent (SE)	
piston	34
observed	30
Atterberg limit	
Wl (%)	32.6
Wp (%)	29.3
Ip (%)	3.3 (4.2 after CBR)
Activity (A)	0.25
WOPM (%)	20.4
γdmax (kN/m$_3$)	18.3
CBR at 95% OPM	13
Swelling (%)	0
γs (kN/m^3)	27.6
γapp (kN/m^3)	13

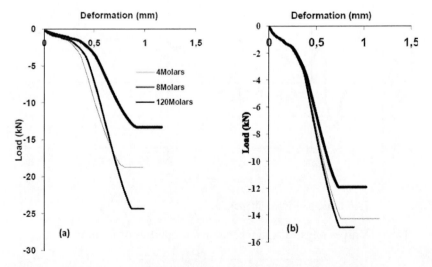

Figure 40. Deformation curve alkali activated schlamm kept at chamber (40°C/60%RH) (view a) and at 80 °C for 28 days (view b).

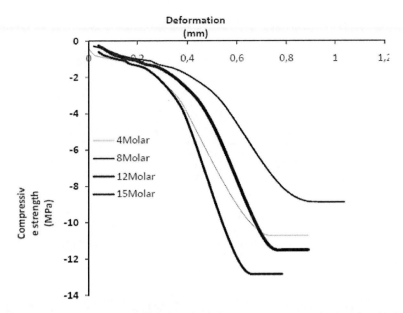

Figure 41. Deformation curve alkali activated feral with 4, 8, 12 and 15 molars NaOH solution and kept in oven at 120 °C for 12hours.

Figure 40 represent the stress–strain curve of Schlamm bricks kept at chamber (40°C/60%RH) and in oven at 80 °C for 28 days and Figure 41 feral activated bricks with 4, 8, 12 and 15 molars NaOH solution kept at oven at 120 °C in sealed Parr type vessels fitted with Teflon liners for 12hours.

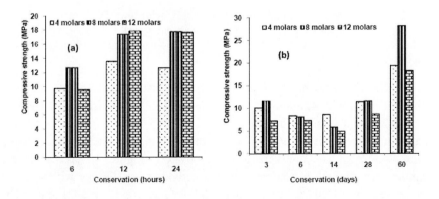

Figure 42. Compressive strength histogram of activated schlamm brick activated with 4, 8 and 12 molars NaOH solution cured at 120°C for 6, 12 and 24hours (view a) and at 40°C/60%RH up to 60 days (view b).

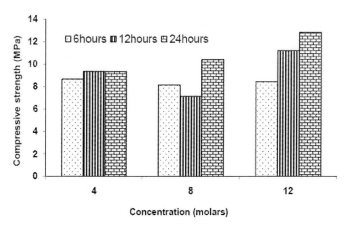

Figure 43. Compressive strength histogram of activated feral bricks with 4, 8, and 12 molars NaOH solution cured at 120°C for 6, 12 and 24 hours.

Figure 44. Views represent typical microstructures of schlamm. View (a) shows the phases present with kaolinite sheets. View (b) represent typical microstructures of schlamm brick samples made with 12 M NaOH solution and cured for 90 days at 40°C and 60°RH.

Figure 42 represents compressive strength histogram of alkali activated schlamm brick with 4, 8 and 12 molars NaOH solution cured at 120°C for 6, 12 and 24hours (view a) and at 40°C/60%RH up to 60 days (view b). The maximum strength is obtained after 60 days when alkali activated schlamm brick with 8 molars are kept at 40°C/60%RH. Figure 43 represents the compressive strength histogram of alkali activated feral bricks with 4, 8, 12 and 15 molars NaOH solution cured at 120°C for 6, 12 and 24hours. The maximum strength is obtained after 24 hours when schlamm brick are alkali activated with 12 molars NaOH solution.

Figure 44 represent the microstructure typical of alkali-activated schlamm brick. The sample was made with 12 M NaOH and cured at 40 °C for 90 days.

The microstructure is more plate-like than the equivalent samples made with 4 and 8 M NaOH, respectively. These latter samples are more granular in appearance that seems to reflect the greater degree of reaction of these samples caused by the higher alkalinity they contain. Figure 45 corresponds to the micrograph of the microstructure of feral treated with 12 M NaOH solution and cured at 120°C for 12 hours.

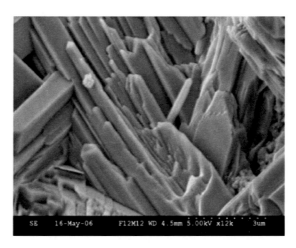

Figure 45. Micrograph represents the microstructure of feral treated with 12 M NaOH solution and cured at 120°C for 12 hours. Based upon X-ray data in Figure 5, it is proposed that these crystals are millisite.

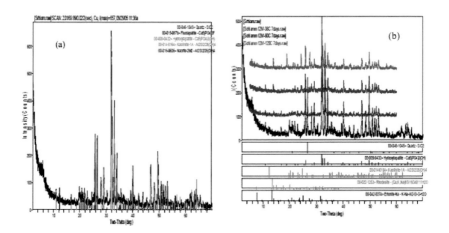

Figure 46. X ray diffraction diagram of starting material Schlamm (view a) and alkali activated product (view b).

Figure 46 represent the x ray diffraction diagram of starting material Schlamm and alkali activated product with 12 molars at 40°C/60%RH, 80°C and 120°C. Figure 47 represent x ray diffraction diagram curves of feral brick alkali activated with 4 (curve 1), 8 (curve 2) 12 (curve 3) and 15(curve 4) molars NaOH solution after 12hours curing. EDS on Schlamm phosphate calcium waste are represented on figure 48. EDS on schlamm sample are coherent with its chemical composition. EDS shows reflect the composition of phosphate mineral: fluorapatite (Ca_5 (PO_4)$_3$ F) or hydroxylapatite ((Ca_5 (PO_4)$_3$ (OH)). Sometimes spectrum corresponds to a transition zone between a clay mineral (kaolinite ($Al_2S_2O_5(OH)$)) and a phosphate mineral. One spectrum corresponds to a quartz mineral covered slightly by a kaolinite mineral. The curing time of the 40° and 80 °C samples was extended to 90 days. Interestingly, the composition of the NaOH solution did affect the outcome of the 90 days runs.

If 4 M NaOH was used, the 80 °C samples tested out to be stronger; if 8 or 12 M NaOH was used, the samples cured at 80 °C tended to be stronger than those cured at 40 °C after 60 days. Figure 47 represent the microstructure typical of alkali-activated schlamm brick. The sample was made with 12 M NaOH and cured at 40 °C for 90 days. The microstructure is more plate-like than the equivalent samples made with 8 and 12 M NaOH, respectively. These latter samples are more granular in appearance that seems to reflect the greater degree of reaction of these samples caused by the higher alkalinity they contain.

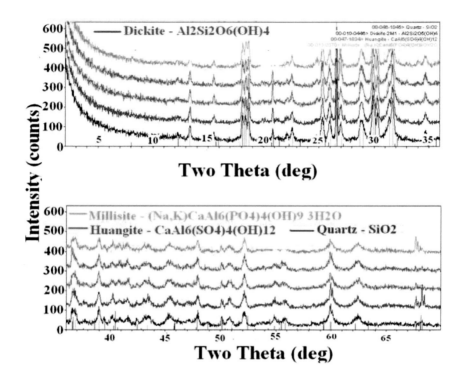

Figure 47. Comparison of starting material feral (curve1) with 4(curve2), 8(curve3), 12(curve4) and 15molar (curve5) concentration reacting with feral after 12hours curing. D = dickite, H = huangite, M = millisite, Q = quartz.

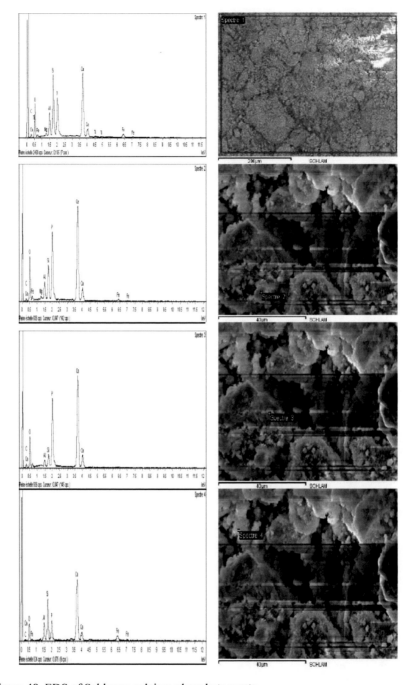

Figure 48. EDS of Schlamm calcium phosphate waste.

A summary of the strength data for schlamm brick is given in figure 42. The curing time of the 40° and 80 °C samples was extended to 90 days. Interestingly, the composition of the NaOH solution did affect the outcome of the 90 days runs. If 4 M NaOH was used, the 80 °C samples tested out to be stronger; if 8 or 12 M NaOH was used, the samples cured at 80 °C tended to be stronger than those cured at 40 °C after 60 days.

Samples fabricated from schlamm and NaOH solutions attain compressive strengths that range from 4 to more than 20 MPa (Figure 42). These strengths are in the range of similar samples made with metakaolin. Strength development is dependent on temperature and length of curing, alkali concentration, fineness, and composition of the raw materials. For samples cured at 120 °C, 12 h temperature seems to impart the best mechanical performance to the samples regardless of the concentration of NaOH solution used to make the sample. It is also notable that the strength gain at any given time is directly proportional to the concentration of NaOH used to make the sample. For reactions carried out at lower temperatures, samples that were cured in the 40 °C/60% RH chamber attained higher strengths than comparable samples cured at 80 °C for the first month.

X-ray diffraction of the schlamm (Figure 46) shows this paragenesis: quartz, kaolinite, fluorapatite, hydroxylapatite and nacrite.

Finally, figure 46 represent the X-ray diffraction patterns for the starting material and the patterns for the 40, 80 and 120 °C samples cured for 7 days made with different molar NaOH solutions. There is not very much difference between the patterns. There is a general broadening of the kaolinite peaks which indicates that some disordered phases are forming that are clay-like, but lack the crystallinity of kaolinite. Traces of erionite are also present. The poorly crystallized phases may be the cementing phases that give the sample its strength.

Figure 45 represents the microstructure typical of alkali-activated feral brick. The sample in the micrograph was made with 12 M NaOH and cured at 120 °C for 12 hours. The micro-structure is more columned-like texture. Figure 49 represent the X-ray diffraction patterns for the 120 °C samples cured for 12 h made with different molar NaOH solutions. There is not very much difference between the patterns. There is a general broadening of peaks with increasing concentrations. Millisite ($(Na, K) CaAl_6(PO_4) 4(OH)_9 3H_2O$) a zeolite phosphate is the most representative of forming minerals; then Huangite ($Ca Al 6(SO)_4(OH)_{12}$). Quartz (SiO_2) and Dickite ($Al_2Si_2O_5 (OH)_4$) are new silicate minerals.

For feral brick Strength (figure 43) development is dependent on temperature and length of curing, alkali concentration, fineness, and composition of the raw materials. The best mechanical performance of the samples depends of the concentration of NaOH solution used to make the sample. For 4 molar concentrations the greatest compressive strength is obtained after 12hours curing. For 8 and 12 molar concentration the greatest compressive strength are obtained after one day curing. The strength given by the Feral treated with 15Molar concentration and cured for 12 hours (8.35 MPa) is less than for the 12Molars concentration and cured for the same duration (12.84 MPa). By using a mixture between "D" sodium silicate and 8Molar NaOH solution (100g 8Molar NaOH + 250g "D" sodium silicate) the compressive strength is neatly increased (16 MPa). This may be due to the fact that with "D" sodium silicate, we have both the formation of zeolite silicates and zeolite phosphates.

After 24 h, the leaching tests of the 120 °C samples feral bricks showed very low conductivities no matter what concentration of alkali was used to make the brick. The values of conductivities increased with time (Table 19), the one day conductivities for all three samples are lower than they are at 7 days.

Table 19. Results of leaching test on schlamm and Feral bricks cured at 120 °C for 12 h Molarity of NaOH used on brick

Concentration (molar)	Sites	Measurement at 1 day		Measurement at 7 days	
		Conductivity (mS/cm)	pH	Conductivity (mS/cm)	pH
4	Schlamm	1.80	10	1.60	10
	Feral	0.7	10	1.00	10
8	Schlamm	4.00	11.5	3.50	11
	Feral	0.6	10	0.90	10
12	Schlamm	6.80	12	5.90	12
	Feral	1.5	11	2.10	11

The 4 M and 8 M samples have the lowest overall conductivities, whereas the 12 M samples are almost twice as high. This suggests that the amount of Na in the 4 and 8 M brick reacts nearly completely, whereas the 12 M sample may contain excess NaOH or soluble sodium silicate which washes out giving it a higher conductivity and pH.

Nevertheless, all conductivity values are reasonably low, proving that reactions are occurring during curing and that zeolite-like mineral(s) are probably forming (Rao and al, 2005). Solubility is low. Durability should be better.

Volcanic Tuffs

The tuff used here to make brick is more or less typical of the materials available to rural persons that make their own brick and then use it to build their homes.

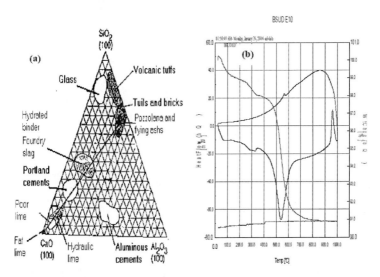

Figure 49. View (a) represents a ternary composition diagram for the system CaO-Al_2O_3-SiO_2 showing the relative position of Bafoundou tuff vis à vis other calcium based cements. View (b) represents the up and down temperature TGA and DTA traces for the tuff, the shape of which resemble heating/cooling curves of kaolinitic clay.

Of the many native soils and tuffs available to work with in Senegal, research to date to develop a more durable brick has focused on one material, an alumino silicate rich tuff (> wt% SiO_2) called Bafoundou that occurs in huge quantities in southeastern Senegal near Kedougou (N'Diaye and al, 2003) See Figure 7. These highly weathered tuffs have traditionally been used as building materials. They are a blend of finely divided kaolinite and quartz (Diop and al, 2006). The tuff is traditionally mixed with water, molded and allowed to dry in the sun. Unfortunately, much like a sun-dried clay clod in a farmer's field, the brick will soften and begin to lose its shape after a few wet-dry cycles. Figure 50 represents a ternary composition diagram for the system CaO-Al_2O_3-SiO_2 showing the relative position of Bafoundou tuff vis à vis other

calcium based cements (view a).:View (b) shows the up and down temperature
TGA and DTA traces for the tuff, the shape of which resemble heating/cooling
curves of kaolinitic clay.

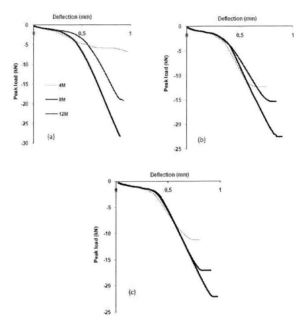

Figure 50. Stress-strain curves for tuff samples cured at 40°C for 28 days (view a),
80°C for 28 days (view b) and 120°C for 12hours (view c). Each sample was made
with 4, 8 and 12molars NaOH solutions as a thick paste and then cured.

Figure 51 shows particle size analysis that suggests that natural sample
and the ultrasonic treated sample are similar (view a). View b represents the
microstructure of the tuffs. The phases present are solid quartz, smaller zeolite
cristals (possibly) and book of kaolinite.

The highly weathered tuffs have important clay content (32%). A
normative study has permitted to determine the mineralogical composition:

Quartz (43%)
Kaolinite (26%)
Illite (6 %)
Albite (5 %)
Ferrous oxide (12%)
Titanium oxide (1%)

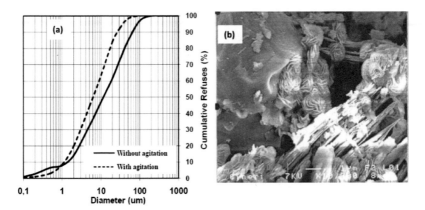

Figure 51. Particle size analysis suggests that natural sample and the ultrasonic treated sample are similar. View (a) represents the microstructure of the tuffs. The phases present are solid quartz (upper left), smaller zeolite cristals (possibly) and book of kaolinite (lower right).

Figure 52 represents the compressive strength histogram of alkali activated tuffs bricks with 4, 8 and 12 molars NaOH solutions cured in sealed Parr type vessels fitted with Teflon liners at 120 °C for 6, 12 and 24hours (view (a)) and with samples kept at chamber (40°C/60%RH) and at oven 80°C for period up to 60days (view (b)). The maximum strength is obtained with 12 molars NaOH concentration with kept at chamber (40°C/60%RH) for 60days.

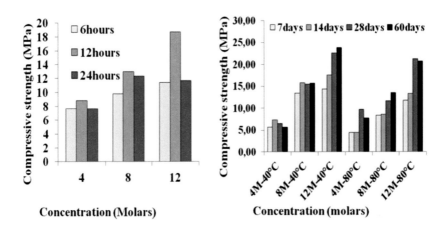

Figure 52. Compressive strength histogram of alkali activated tuffs bricks with 4, 8 and 12 molars NaOH solutions cured in sealed Parr type vessels fitted with Teflon liners at 120 °C for 6, 12 and 24hours (view a) and with samples kept at chamber (view b) at 40°C/60%RH for period up to 60days.

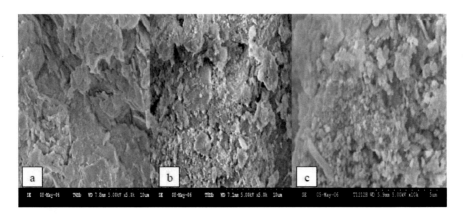

Figure 53. View (a, b, c) represent typical microstructure of tuffs brick samples made
with 4M (view a), 8 M (view b) and 12 M NaOH solution (view c) and cured for 28
days at 40°C.

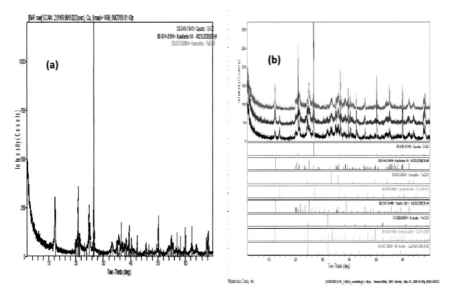

Figure 54. Comparison of starting material Bafoundou tuff (view (a)) and with a 12M
40, 80 and 120°C zeolitized sample (view (b)). The kaolinite and quartz peaks are
smaller and the back ground larger. The pink lines seem to signify the presence of
hydroxysodalite as well as hydrated iron [(Fe$_3$O (OH)].

Table 20 represents the chemical composition of volcanic Bafoundou tuffs samples in weigh percent. Figure 53 represent typical microstructure of tuffs brick samples made with 4M (view a), 8 M (middle) and 12 M NaOH (view b) solutions and cured for 28 days at 40°C/60%RH. Figure 54 shows a comparison of Bafoundou tuff (view a) with a 12M 40, 80 and 120°C zeolitized sample (view b). The kaolinite and quartz peaks are smaller and the back ground larger. The pink lines seem to signify the presence of hydroxysodalite as well as hydrated iron [$Fe_3O(OH)$].

Table 20. Chemical composition
of volcanic Bafoundou tuffs samples (wt %)

SiO_2	Al_2O_3	Fe_2O_3	CaO	MgO	K_2O	Na_2O	TiO_2	Mn_2O_3	P_2O_5	Fire loss	reactive Silica
69.7	13.3	10.1	0.1	0.1	1.3	0.6	0.8	0.1	0.03	4	8.9

Discussion

For samples cured at 120 °C, regardless of the concentration of NaOH used, 12h of curing gives the best compressive strength. This can be explained by the fact that with temperature, the reaction between the alkali solution and the clay has taken place. After 12hours of curing, the hydrated phases that form is most likely zeolitic, because the bulk composition of the starting material falls within a compositional range typical for zeolites. Although the zeolites minerals that form are not as "stable" as kaolinite, they are able to bond to each other and form a solid that is much more resistant to softening and deformation during annual wet/dry cycles. All zeolites are metastable to some degree (Breck, 1974) (Dyer; 1988). The ones that form first, in the presence of abundant water, will become less hydrated and undergo phase transitions as a result of diagenesis. If placed in an aggressive environment such as one with an acidic pH, they will also dissolve. In neutral and alkaline environments however they are very insoluble. Zeolites will change with geological time if buried, but if exposed to heat and humidity at the surface, this change is nearly imperceptible. The type of zeolite that form during the manufacture of the aforementioned brick is somewhat temperature dependant. What will happen on occasion is that an initially formed zeolite will change into another one, one that is more stable. This is a problem associated with nucleation and growth from supersaturated solutions and is similar to what happens during diagenesis. Early formed zeolites are often the least stable converting to a more stable form after a few hours or days of curing. This makes it mandatory for the person making brick using this method (especially at 120 °C) to test strength versus time to see if a disruptive phase change occurs that could reduce performance by introducing shrinkage/expansion cracks. If these occur, samples should not be cured longer than necessary to achieve initial maximum strength.

Generally, for most of the samples tested, those mixed with 12 molar NaOH had the highest strengths. This is what one might expect, because as one increases the concentration of the alkali solution more metakaolin will dissolve in the solution and more sodium aluminosilicate precursors will form that have the ideal ratio to form zeolite A, i.e. Na:Al:Si = 1:1:1). However, increasing the concentration of NaOH indefinitely (e.g. 15 M) will not cause a continuous increase in the strength; rather it will cause more sodium rich phases to form that may not be as insoluble as zeolites. This will increase solubility and possibly have a negative effect on durability.

For samples cured at 40 and 80 °C, the phase and strength evolution varies with each concentration and the method of curing. With 4M concentration, the best strength is obtained when the bricks are cured at 40 °C/ 60% relative humidity (RH) for two weeks. Longer curing does not increase the compressive strength, so curing should not be carried beyond this time. If bricks made with the same concentration are cured in the oven at 80 °C, the best compressive strength is obtained after three weeks.

With 4M concentration, after one or two weeks curing at 40°C/60% RH these samples reach a strength plateau giving the best compressive strength. If one wants greater strength then one should use 80°C curing because the strength tends to increase rather than plateau and longer curing will lead to greater strength.

With 8M concentrations, things appear to be different. Curing at 40°C/60% RH leads to the optimum strength. Strength remains constant after 2 weeks, so under these conditions there is no need to cure bricks more than two weeks. When bricks are cured at 80°C/very low RH, strengths remain constant between1 and2weeks, and then they begin to increase once again.

With 12M NaOH solution, when bricks are cured at 40°C/60% RH, strengths keep increasing up to 60 days. When bricks are cured at 80°C/very low RH, strength keeps increasing until 28 days have passed, then compressive strength remains constant up to 60 days.

The strengths and leachabilities of the samples are similar to those for alkali activated metakaolinite samples which usually run about 3MPa, have a pH of about 10 and a conductivity of 2–3 mS/cm (Bao et al, 2004), (Bao et al,

2005), (Bao and Grutzeck, 2006). The ability to make a brick with these characteristics is rather exciting because of its implications and potential impact on the nature of sun dried brick making in developing countries.

Mostpha et al (2010) used different geopolymer binder preparations based on Saudi Arabia low kaolinitic clay (Table 15). Geopolymer binders activated with NaOH + sodium silicate solutions were prepared as follows; the predetermined amount of NaOH was dissolved in sodium silicate solution then mixed with calcined clays in a porcelain mortar for 5 min. Different geopolymer compositions with various oxide ratios were tested for pure kaolin (KA). Since different kinds of geopolymer clay require different amounts of water, the water was added in the amount necessary to obtain equal consistency. For all clay mixes, the same amounts of NaOH and sodium silicate were used as in Table 16 in order to rich the optimum mix empirically. These mixes empirically optimized because it is hard to quantify the exact reactive aluminum silicate components of clays. It is well known that a 10% relative error is assumed for this semi-quantification method (Moore and Reynolds, 1989). However, the approximate active oxide ratios are given in Table 21 for each mix based on semi-quantitative mineralogical composition of clay raw materials. The total reactive SiO_2 and Al_2O_3 was calculated by correcting for the amounts of unreactive constituents (quartz and feldspars) present clays. The resulting mass was molded into cubic specimens 1x1x1 inch. The molded samples were allowed to mature at room temperature for 24 hours then were cured at different temperatures. The curing temperatures included; (a) room temperature for 3 and 7 days; (b) 85°C for 3 days. After curing, samples are subject to compressive strength test, XRD and SEM investigations. Compressive strength data was obtained from an average of 4 specimens for each mix. The compressive strengths of geopolymer binder samples activated with different alkaline sodium silicate solution and cured at different temperature are given in Table 23. The strength of the reference mix (KA) increases with increasing curing temperature from room temperature to 85°C.

Geopolymer binders prepared from low koalinitic clays (WC and GC) with high Na_2O/Al_2O_3 ratios (mixes 1 and 2) attain no setting after 2 days at both curing temperatures. However, geopolymer binder prepared from RC (RC1) attains no setting after 2 days at both curing temperatures.

The compressive strengths of RC mix with low Na_2O/Al_2O_3 ratio (RC3 and RC4) increase more than the corresponding of the reference mixes at both curing temperatures. Geopolymer prepared from GC show the lowest compressive strength at both curing temperatures.

**Table 21. Semi-quantitative mineralogical composition of clay raw
materials by means of X-ray diffraction analysis**

Phase	Kaolin (KA)	White clay (WC)	Grey clay (GC)	Red clay (RC)
Kaolinite ($Al_2Si_2O_5(OH)_4$)	84	-	32	40
Illite ($K_{0.8}Al_2(Si_3.2Al_{0.8})O_{10}(OH)_2$)	16	-	-	-
Montmorillonite ($Al_2Si_4O11.xH_2O$)	-	45	8	12
Calcite ($CaCO_3$)	-	-	15	-
Quartz (SiO_2)	-	55	15	18
Hematite (Fe_2O_3)	-	-	2	4
Albite ($NaAlSi_3O_8$)	-	-	7	26
Microcline ($KAlSi_3O_8$)	-	-	13	-
Total amount of clay minerals	100	45	40	52

The compressive strengths of RC samples increase more than the reference mixes (KA) at curing temperatures 85°C. The strength difference seems to arise from the presence of sand and feldspars as filler in RC raw material, which form composites with geopolymer matrix. The presence of silica as the main impurities in clays raw materials may play a role in improving the mechanical properties of the geopolymer matrix, as it provides fine particle filers to geopolymer matrix. The silicon and aluminum oxides in the metakaolin reacts with the alkaline liquid to form the geopolymer paste that binds the unreactive constituents associated with clay minerals like quartz and feldspars (albite and microcline) to form the geopolymer composites.

**Table 22. Mix compositions of pure kaolin as well as
of different clays geopolymer**

	Mix proportions (wt %)			Molar oxide ratio	
	Calcined clay	Na-Silicate	NaOH	SiO_2/Al_2O_3	Na_2O/Al_2O_3
KA1	100	121	6.4	4.0	1.0
KA2	100	50.8	021.6	3.	1.0
KA3	100	50.8	5.6	3.0	0.5
KA4	100	28.8	10.0	2.7	0.5
WC1	100	121	6.4	7.5	3.2
WC2	100	50.8	21.6	4.4	3.2
WC3	100	50.8	5.6	4.4	1.6

Table 22. (Continued)

	Mix proportions (wt %)			Molar oxide ratio	
	Calcined clay	Na-Silicate	NaOH	SiO_2/Al_2O_3	Na_2O/Al_2O_3
WC4	100	28.8	10.0	3.4	1.6
GC1	100	121	6.4	11.2	4.3
GC2	100	50.8	21.6	7.5	4.3
GC3	100	50.8	5.6	7.5	2.1
GC4	100	28.8	10.0	6.2	2.1
RC1	100	121	6.4	6.5	2.9
RC2	100	50.8	21.6	3.6	2.9
RC3	100	50.8	5.6	3.6	1.4
RC4	100	28.8	10.0	2.7	1.4

Table 23. Compressive strength (MPa) of different geopolymer mixes hydrated at different temperatures

	Mix	Room temperature		85°C
		3 days	7 days	3days
	KA1	25.6	35.7	43.8
	KA2	42.9	48.8	52.9
KA	KA3	46.6	49.4	51.5
	KA4	40.7	41.7	43.0
	WC1	-	-	-
	WC2	-	-	-
WC	WC3	34.5	37.6	39.4
	WC4	40.3	43.2	46.0
	GC1	-	-	-
	GC2		-	-
GC	GC3	35.4	36.6	37.9
	GC4	29.5	33.7	35.6
	RC1	-	-	-
	RC2	46.5	52.3	59.2
RC	RC3	51.3	57.8	60.1
	RC4	53.5	55.9	63.2

In Portland cement concrete and geopolymer concrete, the coarse and fine aggregates occupy about 75 to 80% of the mass of the concrete. Thus, in designing binders based on low clay raw materials the unreactive constituents (such as quartz, feldspars and hematite) can be considered as a fine aggregate. Many low value clays can be used as feedstocks for geopolymer products.

Variations in the ratio of alumina to silica, and alkali to alumina, produce geopolymers with different physical and mechanical properties.

The presence of calcium carbonate in GC clay may cause the poor mechanical properties of the produced geopolymer due to transformation to CaO with calcination. Zaharak et al. (2009) found that addition of CaO in the initial mixture has a detrimental effect on the final compressive strength in slag-based geopolymer cements. However addition of pulverised silica sand improves strength. Temuujin et al. (2008), found that calcium ion interfere with the polymerization process and alter the microstructure. However, Buchwald et al. (2005) found that, addition of calcium as calcium hydroxide improve mechanical properties of fly ash based geopolymers. In the present study, the presences of Ca^{2+} as CaO in calcined clay decrease the mechanical properties. Thus, the effect of CaO seems to depend on the type of geopolymer. In order to design of geopolymer concrete mixtures based on low kaolinitic clays, the effect of different possible impurities existing in clay must be completely examined.

Conclusion

Zeolites are very insoluble. They are currently forming at the bottom of the World's oceans. However, the Na ions in a zeolite are mobile, which accounts for a zeolite's ability to exchange cations with other substances in solution. In this case the measurement of conductivity used here is actually measuring two things: the degree of reaction that the sample has undergone prior to being tested, and the mobility of the Na ion in the zeolitic-matrix as it exchanges with protons in the water. If the conductivity of the solution is low this suggests that the NaOH that was used to make the sample has reacted with the tuff and has been "tied up" in a tectosilicate matrix. Conductivity reflects the degree of fixation (effectiveness of the recipe to do what it is meant to do) and the magnitude of the cation exchange of Na+ for H_3O^+ that takes place. It is safe to say that if no NaOH had reacted it would dissolve in the leaching solution and conductivity would be in the 20–30 mS/cm range. Because we are in the 1–3 mS/cm range, this suggests that zeolites are forming and they are very much part of the structure even though their presence may not be evident in SEM or X-ray diffraction scans. Based on the low conductivity numbers of 120°C samples, it is predicted that durability of the long term room temperature cured alkali activated tuff brick should be better than that of conventional sun dried clay brick.

The development of a low temperature process to create durable bricks can be accomplished according to the needs of the community. Brick made with this technology is well suited in areas where energy is unavailable; the bricks are cured in the sun. The process does not generate chemical pollutants like fired clay bricks. It can use by-product materials like industrial waste enabling brick makers to solve environmental problems. A wooden mold and a mix of tuff and seawater (if commercially produced silicates are not available) are enough to fabricate good quality block with this technique. Locally available materials can be tested with different amounts of NaOH as mixing solution and cured as a function of temperature to determine the optimum concentrations of NaOH to use. At this time, we recommend 8M NaOH because it seems to be a compromise between strength and cost. If a stronger brick is needed, one can use 12M NaOH because strength was significantly higher in this case. In addition to its value for developing countries, the process has implications all over the world where clean technology is an important consideration.

It seems possible that villagers could start their own businesses providing income and at the same time upgrading the brick used to build houses on a village to village basis. Because the reaction of the tuff with 8M NaOH will continue for a long time at 40 and 80°C, it seems probable that curing could take place under a heavy tarp in the sun, a situation available to villagers all over equatorial Africa, one that will provide and maintain both high humidity and temperature.

It is also proposed that a simple manual press might be used to make full sized bricks. The mixture is quite plastic much like ball clay and can also extruded.

After this study of activating clay to fabricate bricks, we cannot draw a general conclusion. Each clay has its own behavior. With the process used, compressive strengths range from 3 to 28 MPa. The lowest strengths are obtained at 40°C and the best at 80°C. In term of concentration, generally 12 molar give the maximum strength sometimes 8 molars. Keeping bricks at chamber 40°C/60%RH is the cheapest process and sometimes give mechanical performance quite comparable to specimens kept at 80 °C or 120°C. When we have to use alkali activation to make brick, each clay should be studied thoroughly. The resistance obtained with mining phosphate wastes gives strength comparable to those of natural clay.

To obtain higher strength (40-60MPa), the process used by Mostpha et al (2010) is more suitable and can employ low kaolinitic clay.

The choice of the clay is fundamental. The clay should be amorphous and have $SiO_2 + AlO_3 + Na_2O$ greater than 80%.

In the interpretation of the consolidation mechanism we refer explicitly to the theory of zeolitic material synthesis rather than to geopolymers. The major difference between these two possible explanations if that zeolitic material are mainly composed of crystal in closed cage structures, while the geopolymerization consist in the formation of a 3D aluminosilicate amorphous gel/polymer, closer to what authors observe here. However, for geopolymer, the phases observed are badly crystallized and resemble more to a gel. What is however new is that currently people who make geopolymer generally use metakaolin which is strongly reactive with soda and sodium silicates. Hence further detailed studies and characterization are necessary to identify exactly the new forming phases.

The durability of the alkali-activated bricks is based on the assumption that zeolites form. However, they are no strong evidence for the presence of such zeolites in the bricks. Having the chemical composition of their mixes falling in the zeolite zone on the SiO_2-Al_2O_3-Na_2O phase diagram does not mean that zeolites have to form regardless of the thermodynamics and the kinetics of the system. There are no thermodynamic calculations showing that a zeolite phase can form at 40°C or 120°C from the corresponding oxides. We have to made additional analysis to know exactly what form (zeolite or geopolymer) to understand and improve the processes though both of them constitute binders.

References

Bao Y, Kwan S, Siemer DD, Grutzeck MW. Binders for radioactive waste forms made from pretreated calcined sodium bearing waste (SBW). *J. Mater. Sci.* 2004;39 (2):481–8.

Bao Y, Grutzeck MW. Solidification of sodium bearing waste using hydroceramic and portland cement binders. Ceram Trans 2005;168:243–52. *Environmental Issues and Waste Management Technologies in the Ceramic and Nuclear Industries X.*

Bao Y, Grutzeck MW, Jantzen CM. Preparation and properties of hydroceramic waste forms made with simulated Hanford low-activity waste. *J. Amer. Ceram. Soc.* 2005;88(12):3287–302.

Bao Y, Grutzeck MW. General recipe and properties of a four inch hydroceramic waste form. In: Ceramic Transactions, vol. 176 (Environmental Issues and Waste Management Technologies in the Ceramic and Nuclear Industries XI), Am Ceram Soc, Westerville, OH; 2006. p. 63–74.

Berg LG, Remiznikova VI, Vlasov VV. Reaction of kaolinite with caustic soda. *KhimiyaiKhimicheskaya Teknologiya* 1965;8(2):181–5.

Berg LG, Remiznikova VI, Vlasov VV. Nature of the interaction of kaolinite with some bases

Bergaya, F., Theng, B.K.G., Lagaly G., 2006. *Handbook of clay science*, Elsevier, Amsterdam.

Billong, N., Melo, U.C., Louvet, F., Njopwouo, D., 2009. Properties of compressed lateritic soil stabilized with a burnt clay–lime binder: Effect of mixture components. *Construction and Building Materials* 23, 2457-2460.

Breck DW. Zeolite molecular sieves. New York: John Wiley and Sons; 1974.

Brew, D.R.M., MacKenzie, K.J.D., 2007. Geopolymer synthesis using silica fume and sodium aluminate. *J. Mater. Sci.* 42, 3990–3993.

Buchwald, A., Dombrowski, K., Weil, M., 2005. The influence of calcium content on the performance of geopolymeric binder especially the resistance against acids, 4th International conference on Geopolymers, 29.6.-1.07.05, St.Quentin, France

Buchwald, A., Hohmann, M., Posern, K., Brendler, E., 2009. The suitability of thermally activated illite/smectite clay as raw material for geopolymer binders, *Applied Clay Science* 46, 300–304.

Castellin, L., Aldon, J., Olivier-Foucarde, J., Juma, J.P., Bonnet, Blanchart, P., 2002. Fe Mösbauer study of iron distribution in kaolin raw material: influence of the temperature and the heating rate. *Journal of the European Ceramic Society* 22(11), 1767–1773.

Congress Geopolymer, Saint Quentin, France, 28 June–1 July, pp. 9–15.

Davidovits, J., 1999. Method for bonding fiber reinforcement on concrete and steel structures and resultant products. US Patent No. 5925449.

Davidovits, J., 2005. Geopolymer chemistry and sustainable Development. The Poly(sialate) terminology : a very useful and simple model for the promotion and understanding of green-chemistry. In: Davidovits, J. (Ed.), Proceedings of the World.

Diop B, Jauberthie R, Melinge Y. Bouguerra. Influence of volcanic tuff fillers on the durability of mortar stored in sulfatic environment. In: Proceedings International symposium, Durability of concrete; Canmet/ACI congress, Montreal, Canada; June 2006.

Davidovits J. Geopolymers and geopolymeric materials. *J. Therm. Anal.* 1989;34(2):429–41.

Davidovits, J., Buzzi, L., Rocher, P., Gimeno, D., Marini, C., Tocco, S., 1999. Geopolymeric cement based on low cost geologic materials. Results from the european research project geocistem. In: Davidovits, J., Davidovits,

R., James, C. (Eds.), Proceedings of the 2nd International Conference on Geopolymer '99, Saint Qunentin, France, June 30–July 2, pp. 83–96.

Davidovits, J., Comrie, D.C., Paterson, J.H., Ritcey, D.J., 1990. Geopolymeric concretes for environmental protection. *Concrete International*;12, 30–39.

De Silva, P., Sagoe-Crenstil, K., Sirivivatnanon, V., 2007. Kinetics of geopolymerization: Role of Al_2O_3 and SiO_2. *Cement and Concrete Research* 37, 512–518.

Diop M.B, Grutzeck M.W., 2008. Low temperature process to create brick. *Construction and Building Materials* 22, 1114-1121

Dombrowski, T., 2000. The origins of kaolinite. Implications for utilization. In: W.M. Carty and C.W. Sinton, Editors, Science of Whitewares II, American Ceramic Society, Westerville, OH, pp. 3–12. Farmer V.C., 1974. in: V.C. Farmer (Ed.), Infrared Spectra of Minerals, Mineralogical Society, London, UK, pp. 331.

Farmer V.C., 2000. Transverse and longitudinal crystal modes associated with OH stretching vibrations in single crystals of kaolinite and dickite, *Spectrochim. Acta* A 56, 463 927-930

Freidin, C., 2007. Cementless pressed blocks from waste products of coal-firing power station. *Construction and Building Materials* 21, 12-18.

Gallagher, P.K., Brown, M.E., Kemp, R.B., 2003. In: Brown, M.E., Gallagher, P.K., Editors, Handbook of Thermal Analysis and Calorimetry: Applications to inorganic and miscellaneous materials, Elsevier.

Grim, R., 1968. *Clay Mineralogy*, McGraw-Hill, New York.

Grutzeck MW, Kwan S, DiCola M. Zeolite formation in alkali-activated cementitious systems. *Cem. Concrete Res.* 2004;34(6):949–55.

Grutzeck MW, Siemer DD. Zeolites synthetized from ClassF fly ash and sodium aluminate slurry.*J. Amer. Ceram. Soc.* 1997;80(9):2449–53.

Gurhan, G., Yoldas, S., Kusoglu M., Kadir I.Y., 2006. Equilibrium and kinetics for the sorption of promethazine hydrochloride onto K10 montmorillonite. *Journal of colloid and interface science* 299, 155-162.

Hos, J.P., McCormick P.G. and Byrne L.T., 2002. investigating of a synthetic He, C., Osbaeck, B., Makovicky, E., 1995. Pozzolanic reactions of six principle clay minerals: activation, reactivity assessments and technological effects. *Cem. Concr. Res.* 25(8),1691–702.

Kakali, G., Perraki, T., Tsivilis, S., Badogiannis, E., 2001. Thermal treatment of kaolin: the effect of mineralogy on the pozzolanic activity, *Appl. Clay Sci.* 20, 73–80.

Kaloumenou, M., Badogiannis, E., Tsivilis, S., Kakali, G., 1999. Effect of the kaolin particle size on the pozzolanic behaviour of the metakaolinite produced. *J. Therm. Anal. Calorim.* 56, 901–907 .

Klinkenberg, M., Dohrmann, R., Kaufhold, S., Stanjek, H., 2006. A new method for identifying Wyoming bentonite by ATR-FTIR. *Applied Clay Science* 33, 195–206.

Komnitsas, K., Zaharaki, D., Perdikatsis, V., 2007. Geopolymerization of low calcium ferronickel slags. *Journal of Materials Science* 42, 3073–3082.

Moore, D.M., Reynolds Jr. R.C., (Editors), 1989. X-Ray diffraction and the Identification and Analysis of Clay Minerals, Oxford Univ.Press.

Madejová, J., Janek, M., Komadel, P., Herbert, H.-J., Moog, H.C., 2002. FTIR analyses of water in MX-80 bentonite compacted from high salinary salt solution systems, *Appl. Clay Sci.* 20, 255–271.

Malek, Z., Balek, V., Garfinkel-Shweky, D., Yariv, S., 1997. The study of the dehydration and dehydroxylation of smectites by emanation thermal analysis, *Journal of Thermal Analysis and Calorimetry* 1(48), 83–92.

Mostafa, N.Y., El-Hemaly, S. A. S., Al-Wakeel, E. I., El-Korashy, S. A., Brown, P.W., 2001. Characterization and Evaluation the Pozzolanic Activity of Egyptian Industrial By-products: I- Silica fume and Dealuminated kaolin. *Cem. Concr. Res.* 31, 467-474.

Mostafa N. Y. and Mohsen (2010), Q. Investigating the possibility of utilizing low kaolinic clays in production of geopolymer bricks, *Ceramic silica* 54 (2) 160-168

Nastro, V., Vuono, D., Guzzo, M., Niceforo, G., Bruno, I., De Luca, P., 2006. Characterisation of raw materials for production of ceramics. *Journal of Thermal Analysis and Calorimet*ry 84 (1), 181–184.

N'Diaye M., Diop M. B, Ngom P. M., Besson J. C. Pouzolanic activity of acid and intermediary volcanic tuffs of Mako areas (Senegal). In: International Symposium on Industrial Minerals and Building Stones, International Association of Engineering Geology and the Environment (IAEGE), Proceedings; September 15–18 2003. p. 517– 23.

Palomo, A., Grutzec, M.W., Blanco, M.T., 1999. Alkali – activated fly ashes. A cement for the future, *Cem. Con. Res.* 29, 1323-1329.

Palomo A, Blanco MT, Granizo ML, Puertas F, Vasquez T, Grutzeck MW. Chemical stability of cementious materials based on metakaolin. *Cem. Concrete Re*s. 1999; 29(7):997–1004.

Panias, D., Giannopoulou, I., Perraki, T., 2007. Effect of synthesis parameters on the mechanical properties of fly ash-based geopolymers. *Colloids and Surfaces A*: *Physicochemical Engineering Aspects* 301, 246–254.

Phair, J.W., Van Deventer, J.S.J,. 2001. Effect of silicate activator pH on the leaching and material characteristics of waste-based geopolymers. *Miner. Eng.* 14(3), 289–304.

Phair, J.W., Van Deventer, J.S.J., 2002. Effect of the silicate activator pH on the microstructural characteristics of waste-based geopolymers. *Int. J. Miner. Process* 66:121–143.

Pichon H. Characterization and quantification ofthe reactive fraction in volcanic pozzolans. Bull de Liaison des Laboratoires des Ponts et Chausse´es 1996;201:29–38.

Provis, J.L., Lukey, G.C., van Deventer, J.S.J. 2005. Do geopolymers actually contain nanocrystalline zeolites? - A reexamination of existing results, *Chem. Mater.* 17, 3075533

Rahier, H., Wastiels, J., Biesemans, M., Willlem, R., Van Assche, G., Van Mele, B., 2007. Reaction Mechanism, kinetics and high temperature transformations. *J. Mater. Sci.* 42, 2982-2996.

Rahier H, Wullaert B, Van Mele B. Influence of the degree of dehydroxylation of Kaolinite on the properties of aluminosilicate glasses. *J. Therm. Anal. Calorimetry,* 2000; 62(2):417–27.

Russell, J.D., Fraser, A.R., 1994. in: M.J. Wilson (Ed.), Clay Mineralogy: Spectroscopic and Chemical Determinative Methods, Chapman and Hall, London, UK, p. 11.

Richardson CK, Markuszewski R, Durham KS, Bluhm, DD. Effect of caustic and microwave treatment on clay minerals associated with coal. In: ACS Symposium Series, vol. 301, *Mineral Matter Ash. Coal*; 1986. p. 513–23.

Sazhin VS, Pankeeva NE. Interaction of kaolinite with sodium hydroxide solutions. *Ukrains'kii Khemichnii Zhurnal* 1967; 33(5): 528–30.

Saikia, N.J., Bharali, D.J., Sengupta, P., Bordoloi, D., Goswamee, R.L., Saikia, P.C., Borthakur, P.C., 2003. Characterization, beneficiation and utilization of a kaolinite clay from Assam. *India. Appl. Clay Sci.* 24, 93–103.

Siemer DD, Grutzeck MW, Scheetz BE. Comparison of materials for making hydroceramic waste forms. Ceram Trans 2000;107:161–7. Environmental Issues and Waste Management Technologies in the Ceramic and Nuclear Industries V.

Slavik, R., Bednarik, V., Vondruska, M., Skoba, O., Hanzlicek, T., 2005. Chemical indicator of geopolymer. Geopolymer 2005 World Congress, pp. 17–19. Saint Quentin. France:

Smykatz-Kloss, W., 1974. Differential Thermal Analysis: Application and Results in Mineralogy. Springer-Verlag, Berlin, eidelberg, New York.

Soro, N.S., Blanchart, P., Bonnet, J.P., Gaillard J.M., Huger, M., Touré, A., 2005. Sintering of kaolin in presence of ferric compound: study by ultrasonic echography, *Journal of Physics* IV 123, 131–135.

Temuujin J., van Riessen A., Williams R., 2008. Influence of calcium compounds on the mechanical properties of fly ash geopolymer pastes. *Journal of Hazardous Materials*, 167(1-3), 82-88.

van Jaarsveld, J.G.S, van Deventer, J.S.J. 1997. The potential use of geopolymeric materials to immobilize toxic metals. I. Theory and applications. *Miner. Eng.* 10, 659– 669.

Van der Marel, H.W., Beutelspacher, H., 1976. Atlas of Infrared Spectroscopy of Clay Minerals and their Admixtures. Elsevier, Amsterdam.

Varlamov, V.P., Kroichuk, L.A., Toporkova, A.A., 1976. A new method for estimating the drying sensitivity of clay. *CeramurgiaInternational* 2(2), 98-101.

Xu Hua, Van Deventer J.S.J., 2000. The geopolymerisation of alumino-silicate minerals. *Int. J. Miner. Process.* 59 (3), 247–266.

Xu H., Van Deventer J.S.J., 2002. Geopolymerisation of multiple minerals. *Miner. Eng.*15, 1131–1139.

Yang, K.H., Song, J.K., Ashour, A.F, Lee, E.T., 2008. Properties of cementless mortars activated by sodium silicate. *Construction and Building Materials* 22, 1981-1989.

Yang, N., Zeng, Y., 2005. Research and development of chemically-activated cementing materials, necessity and feasibility [C] Zeng Y., Fang Y., Xu L. Research Progress in Chemically-Activated Cementing Materials. Nanjing: Southeast University Press 1-14.

Yip, C.K., Lukey, G.C., Provis J.L., van Deventer, J.S.J., 2008. Effect of calcium silicate sources on geopolymerisation. *Cem. Con. Res.* 38(4), 554-564.

Zaharaki, D., Komnitsas, K., 2009. Effect of additives on the compressive strength of 586 slag-based inorg.

Chapter 8

NEW IDEAS

Clays are extremely abundant minerals and exist in an endless variety with extremely different characteristics. From a pragmatic point of view for construction purposes, two main families have to be considered: swelling clays (e.g. smectites, montmorillonite, etc.) or non swelling, or very lightly swelling clays clays (e.g. kaolinites, illites, chlorites, etc..).Very often, binder part of the earth contains several type of clay. The cohesion of the clay containing binder is strongly dependant on the clay type. Non swelling clays (e.g. kaolinites,) confer little cohesion, resulting in weak components and structures. On the other hands, swelling clays (e.g. smectites, etc...) confer high cohesion provided that they contains high valence interlayer ions, but the initial shrinkage after manufacturing of the component will result in excessive cracking.

In north of Senegal in a village called Tiago, we find clay that used people to make their bricks which is not sensitive to water. After raining season, no changes are noted on constructions. It would be interesting to inventory and study all such clays

An interesting idea to develop is to study the earth of termites which rural person uses to make strong habitation in south eastern Senegal. The substance released by these organism sticks particle each other and enable to make strong waterproof bricks.

A crucial point to improve the method of alkali activation is the dissolution of clay mineral. The Si-O bonds are strong. In the alkali activation process only part of the clay are dissolve this is on of the reason that explain we face difficulties to have easily mechanical performance comparable to those of Portland cement. A method to destroy the Si-O bonds is to keep clay sample in a humidity chamber 90°C/60% relative humidity like the weathering and deterioration of alkali silicate glasses by humidity, a common problem. The weathering and corrosion mechanisms, especially in the case of soda-lime-silica float glass are well understood. The physical and chemical adsorptions of water on the surface are the first steps in the reaction. The presence of unsatisfied network bond and undercoordinated sodium and calcium ions on the surface provide the primary driving forces for physical adsorption water. The subsequent chemisorptions of water to form surface silanols is facilitated by non bridging oxygen at the surface and in the bulk of the glass. In figure 1, is represented melt and polished surfaces prepared, then subjected to elevated temperature and humidity. Samples were weathered in a static temperature and humidity weathering chamber. The temperature was controlled at 60°C while the humidity was controlled at 90% relative humidity. The samples were preheated to 90°C prior to entering the humid environment to prevent condensation. These images (figure 1) were obtained in secondary electron mode at magnifications between 500X and 1,500X, and illustrate the extent of weathering-related surface corrosion. We notice the difference and the evolution of an alkali silicate glass in a humidity chamber (Fig. 1).

The formation of silanol renders the surface more polar and further enhances adsorption, even condensation of water on the surface. Temperature/dew point fluctuations and gradient as well as capillarity effects due the close proximity of surfaces can further enhance the condensation of water films on the surface. The presence of an aqueous liquid layer on the surface, albeit nanoscopic, in thickness, is especially important, because it enhance the kinetic of in-depth hydration by ion-exchange between protons/hydronium-ions in the liquid layer and sodium and calcium ions in the glass.

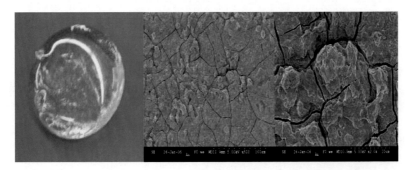

Figure 1. Samples were weathered in a static temperature and humidity weathering chamber.

The net effect can be a dramatic increase in the alkali activity (i.e., pH) of the surface because of the out-diffusion of sodium and calcium into the aqueous liquid surface layer. If the alkalinity of this surface layer can be maintained below ~pH9, leaching and carbonate formation can occur. However, once the alkali activity reaches a critical value, corresponding to pH $\geq$ 9, the liquid surface layers attacks the silicate network, which causes it to dissolve locally.

To create this atmosphere is easily done in hot countries like California or Senegal in West Africa where the temperature can reach 45-50°C.

Last, we know that most of sedimentary rocks (carbonates, ferrous rocks, phosphates etc...) are forming at ambient temperature at the surface of the earth crust or in the ocean. For example, it's well known that when released by the processes of deterioration of the rocks, Ca enters in solution and water transport it in the form of $(CaH_2 (CO_3)_2)$ ionized (bicarbonate or acid carbonate Ca).

It is brought by the rivers to the oceans. The differences between fresh water and sea water are very marked. Thus in the first the report/ratio Na/Ca is of 0.28, while in the seconds, it is of 26.63, that is to say 100 times higher. This calcium fall at sea is due to the behaviour of Ca^{2+} in solution. A weak increase in the pH is enough to cause its precipitation in the form of $CaCO_3$. The current sediments would contain on average 12.5% of CaO, according to Kuenen.

The annual contribution of the rivers is estimated by Clark at 6 billion tons of $CaCO_3$; this quantity is equal to that which settles each year on the sea-bed. That means that the sea water is saturated with $CaCO_3$.

This release of ions of weak acid (HCO_3) and strong cations (Ca^{2+}) gives to sea water a low alkalinity (pH~8).

It is thus the presence of carbonic gas which maintains Ca^{2+} in solution. However its solubility varies with the temperature of water. It is much stronger in the cool water (double) that in warm water. The algae which absorb CO_2 and raise the pH contribute to the precipitation of $CaCO_3$. However, the development of the algae depends on the content of water phosphates and out of nitrates; these substances are especially abundant on the continental platform. Calcareous sedimentation will be thus especially the prerogative of warm water and not very deep.

The idea of using this natural process to make bricks is to raise the pH of sea water to contribute to the precipitation of $CaCO_3$.

REFERENCES

Breck D. W. 1974. Zeolite molecular sieves, structure, chemistry and use, New York: John Wiley and Sons; New York, a Wiley Interscience Publication.

Diop Mouhamadou B., Robert Schaut, Nicholas Smith, Carlo Pantano, "Weathering of boron Oxide-Doped Soda-Lime-Silicate Glass surfaces", Glass and Optical Materials division spring 2006 Meeting, May 16-19, 2006 Hyatt Regency Greenville, South Carolina USA, Site web: http://www.ceramics.org/gomd06.

Schaut A and Carlo g Pantano Acid interleave coatings inhibit float-glass weathering, corrosion Robert.

Kuenen, Ph. H. (1956): Classification of *carbonate rocks.* - American Association of Petroleum Geologists, Journal of Sedimentary Petrology, 22, 64-72...

Chapter 9

CONCLUSION

With alkali activation of clay we notice that we can reach compressive strength around 60 MPa easily. The crucial questions that arise are;

Can we reach performances obtained with Portland cement?
How can we do it?
At what cost?.

The answers of all these questions are positively stated below.

In order to reach performances obtained with Portland cement using alkali activation of clay, we have to pre-treat the clay. However,weshould precise that instead of heating the material, we just need to keep it at humidity weathering chamber levelled around 60°C and at 90% relative humidity to partially destroy the strong Si-O bonds.

The second step would be to optimize the mix such as it is done with Portland cement. Note that we should cautisouly pay attention to the importance of estimating the correct quantity of clay, water and caustic acid to be used. In reality, it is hard to to quantify the exact reactive aluminium silicate components of clays. According to Moore and Reynolds (1989), it is well known that a 10% relative error is assumed for this semi-quantification method.

A more compelling difficulty is to decide between using caustic solution or solid grains. Since the reaction of caustic and water is very exothermic while being a catalyst for the reaction between clay, water and caustic. I have figured out that it is most convenient to use solid caustic grain (NaOH and/or Sodium silicate) in the process. Depending on the composition of the raw clay-either, SiO_2, Al_2O_3, Na_2O and CaO and P_2O_5 for phosphate waste, we will determine the choice of caustic. Ideally, the clay,the clay should have the the sum $SiO_2 + Al_2O_3 + Na_2O + CaO$ greater than 80% in weight because these oxides are the most involved in reactions.

Preferably, the clays should not be crystalline. The moulded samples would be allowed to mature at room temperature for 24 to 48 hours then cured at different temperatures. The curing temperatures should include: (a) room temperature (40°), (b) 85°C and (c) 100-120°C. The only reliable method to know exactly the duration of conservation of samples will be known by tests.

The problem of costs is very important. Clay is the most abundant rock of the earth crust and generally easily available almost everywhere and workable. The fact that we can use by-products is also a mean of lowering costs on hand and the possibility of replacing caustic by seawater, on the other hand.

An important advantage of alkali activated bricks process is that it does not generate pollutants and can assist with environmental problems.

Last but not least, we see that the idea of using natural process taking place on the earth crust will involve undoubtedly the collaboration of chemist, geologist and material scientist. I am convinced that the progress of humanity in science and technology and to face challenges is to observe well and take example on the nature.

ACKNOWLEDGMENTS

Most of the results presented in this book on zeolites/gepolymer binders came from my stay at Penn state university. From September 2005 to May 2006, I received a Senior Fulbright scholarship from the US government.

I am very grateful to Professor Kwado Osseo Asare, Distinguished Professor of Metallurgy and Energy and Geo-environmental Engineering at Penn State University who did recommend my application to Professor Carlo G. Pantano and Professor Michael G. Grutzeck.

Professor Carlo G. Pantano, Distinguished Professor of Materials Science and Engineering, Director of the Materials Research Institute (MRI) at Penn State University welcomed me in his research team and gave me all the necessary support to make my research successful.

Prof Michael G. Grutzeck, Professor of Materials at Penn State University initiated me to zeolites/gepolymer binders. Also, he developed friendly relationships, which helped me support the American climate.

I am very grateful to Professor Farshad Rajabipour, Energy and Mineral Engineering Department at Materials Research Institute of The Pennsylvania State University who kindly wrote the preface of this book.

Last but not least, I am grateful to Professor El hadj Kandji of University Cheikh Anta Diop and Miss Soukail, Ph. D who helped in correcting syntax and grammar of the book.

ANNEXE 1

SODIUM HYDROXIDE

Sodium hydroxide (NaOH), also known as lye or caustic soda, is a caustic metallic base used in industry, mostly as a strong chemical base in the manufacture of paper, textiles, and detergents.

Sodium hydroxide	
General	
Systematic name	Sodium hydroxide
Other names	Lye, caustic soda
Molecular formula	NaOH
Molar mass	40.0 g/mol
Appearance	White flakes
CAS number	[1310-73-2]
Properties	
Density and phase	2.1 g/cm^3, solid
Solubility in water	111 g/100 ml (20 °C)
Melting point	323 °C (596 K)
Boiling point	1390 °C (1663 K)
Basicity (pK_b)	0.2
Hazards	
MSDS	External MSDS

(Continued)

Sodium hydroxide	
Hazards *continued*	
EU classification	Corrosive (C)
R-phrases	R35
S-phrases	S1/2, S26, S37/39, S45
Flash point	non flammable
Supplementary data page	
Structure andproperties	n, ε_r, etc.
Thermodynamic data	Phase behavior Solid, liquid, gas
Spectral data	UV, IR, NMR, MS
Related compounds	
Other anions	sodium chloride, sodium sulfate
Other cations	potassium hydroxide, calcium hydroxide
Related bases	ammonia lime
Related compounds	chlorine

General Properties

When pure, it is a white solid, available in pellets, flakes, granules, and also 50% saturated solution. It is very deliquescent and readily absorbs carbon dioxide from the air, so it should be stored in an airtight container. It is very soluble in water (with liberation of heat), ethanol, and glycerin. It is insoluble in ether and other non-polar solvents. Sodium hydroxide is produced in the chloralkali process, which is the electrolysis of an aqueous solution of sodium chloride. It is a by-product from the production of chlorine. A sodium hydroxide solution will leave a yellow stain on fabric and paper.

Nomenclature

Both sodium hydroxide (NaOH) and potassium hydroxide (KOH) are commonly called "lye" in North America, which can lead to some confusion.

However, most commercially available lye is NaOH. Lye is also a main ingredient in the making of soap. NaOH is now most commonly used for this, but traditionally KOH was used because it was easier to obtain.

Uses

General Applications
Sodium hydroxide is the principal strong base used in the chemical industry. In bulk it is most often handled as an aqueous solution, since solutions are cheaper and easier to handle. It is used to drive for chemical reactions and also for the neutralisation of acidic materials.

Soap Making
Soap making via saponification is the most traditional chemical process using sodium hydroxide. The Arabs began producing soap in this way in the 7th century, and the same basic process is still used today.

Biodiesel
For the manufacture of biodiesel, sodium hydroxide is used as a catalyst for the transesterification of . This only works with anhydrous sodium hydroxide, because water and lye would turn biodiesel into soap. It is used more often than potassium hydroxide because it costs less, and a smaller quantity is needed for the same results. Another alternative is sodium silicate.

Food Preparation
Food uses of lye include washing or chemical peeling of fruits and vegetables, chocolate and cocoa processing, caramel color production, poultry scalding, soft drink processing, and thickening ice cream. Olives are often soaked in lye to soften them, while pretzels and German lye rolls are glazed with a lye solution before baking to make them crisp.

Lye is used to make the Scandinavian delicacy known as lutefisk (from *lutfisk*, "lye fish"). Hominy is dried maize (corn) kernels reconstituted by soaking in lye-water.

Domestic Uses
Sodium hydroxide is occasionally used in the home as an agent for unblocking drains, but it is highly caustic and should be handled with care (see precautions).

Precautions

Gloves, eye protection should be worn when using sodium hydroxide, since there is a high danger of causing chemical burns, permanent injury or scarring, and blindness. A PVC apron is also recommended when concentrated solutions or the solid form are used. It should be stored well away from strong acids such as battery acid. It can create enough heat to ignite flammables (such as alcohols), so it should be added slowly in biodiesel processors. Vinegar is a mild acid that will neutralize lye if it were to make contact with the skin.

References

N. N. Greenwood, A. Earnshaw, *Chemistry of the Elements*, 2nd ed., Butterworth-Heinemann, Oxford, UK, 1997.
Heaton, A. (1996) *An Introduction to Industrial Chemistry*, 3rd edition, New York:Blackie. ISBN 0-7514-0272-9.

SODIUM SILICATE

Sodium silicate, also known as water glass, is a compound used in cements, textile and lumber processing.

Sodium metasilicate	
General	
Other names	Waterglass
Molecular formula	Na_2SiO_3
Molar mass	122.06 g/mol
Appearance	colorless solid
CAS number	[6834-92-0]
Properties	
Density and phase	2.4 g/cm^3, solid
Solubility in water	Soluble
Melting point	1088 °C

Boiling point	? °C (? K)
Sodium metasilicate	
Thermodynamic data	
Standard enthalpy of formation $\Delta_f H°_{solid}$	−1519 kJ/mol
Standard molar entropy $S°_{solid}$	113.8 J.K^{-1}.mol^{-1}
Hazards	
EU classification	not listed
NFPA 704	
Supplementary data page	
Structure and properties	n, ε_r, etc.
Thermodynamic data	Phase behavior Solid, liquid, gas
Spectral data	UV, IR, NMR, MS
Regulatory data	Flash point, RTECS number, etc.
Related compounds	
Other anions	Sodium carbonate, Sodium germanate, Sodium stannate, Sodium plumbate
Other cations	Potassium silicate

Properties

Sodium carbonate and silicon dioxide react when molten to form sodium silicate and carbon dioxide.

Sodium silicate is a white solid that is soluble in water, producing an alkaline solution. There are many kinds of this compound, including sodium orthosilicate, Na_4SiO_4; sodium metasilicate, Na_2SiO_3; sodium polysilicate, $(Na_2SiO_3)n$; sodium pyrosilicate, $Na_6Si_2O_7$, and others. All are glassy, colourless and dissolve in water.

Sodium silicate is stable in neutral and alkaline solutions. In acidic solutions, the silicate ion reacts with hydrogen ions to form silicic acid, which when heated and roasted forms silica gel, a hard, glassy substance that absorbs water readily.

Metal Repair

Sodium silicate is used, along with magnesium silicate in muffler repair paste. When dissolved in water, both sodium silicate, and magnesium silicate form a thick paste that is easy to apply. When the exhaust system of an internal combustion engine heats up to its operating temperature, the heat drives out all of the excess water from the paste. The Silicate compounds that are left over have glass-like properties, making a somewhat permanent, brittle repair.

Food Preservation

Sodium silicate was also used as an egg preservation agent in the early 20th Century with large success. When fresh eggs are immersed in it, bacteria which cause the eggs to spoil are kept out and water is kept in. Eggs can be kept fresh using this method for up to nine months.

Timber Treatment

The use of sodium silicate as a timber treatment for pressure-treated wood began in 2005, after an environmental chemist's research on allergies and autism branched into her developing a method for rendering sodium silicate insoluable once the lumber has been treated with it. This treatment preserves wood from moisture and insects and possesses some flame-retardant properties. Sodum silicate treated lumber is considered a safer alternative to chromated copper arsenate (CCA, restricted by the EPA in 2004) and alkaline copper quaternary (ACQ, which corrodes non-galvanized nails and screws).

References

International Chemical Safety Card 1137.
Texas AandM University: Simplified food-oil refining.
Timber Treatment Technology's chart of sodium silicate applications, including wood preservation.

ANNEXE 2

LOW TEMPERATURE ALKALI ACTIVATED SILICATE BLOCK

Invention Disclosure

February 14, 2006

Introduction

Earthen block, cement brick, hydraulic lime brick and fired clay brick are examples of building materials used to build houses in developing countries. All these technologies have notable disadvantages that are solved by the new invention disclosed here.

Earthen block are simple to make from readily available natural materials, but earthen block do not resist water immersion because of their clay content. The dry clay becomes wet which leads to softening and the development of cracks that destroys the block after few months. Cement brick entails the use of Portland cement, which is the primary CO_2 producing industry in the world and is expensive and thus does not represent a solution for developing countries.

Hydraulic lime entails the firing of limestone at 1000 °C, which is also expensive, some times more so than Portland cement. A fired clay brick is made in a high temperature wood, gasoline or electrically heated kiln, and consumes considerable energy. The disclosed invention provides advantages vis à vis these technologies.

For example, alkali activated silicate block are more durable than earthen block, but they can be made using almost identical processing techniques. They are more cost efficient than cement brick, hydraulic lime brick, and fired clay brick because they can be made at near ambient temperatures well below the boiling point of water.

The making of alkali activated silicate brick can function at small scale or at industrial scale according to the needs of the community. Brick made with this technology can be obtained only after few hours (at elevated temperatures) or after three to for weeks in an area where energy is unavailable and samples are cured in the sun.

The process is less expensive because it requires only a silicate material e.g. natural soil or rock or even artificial material like industrial waste and a concentrated caustic solution.

A mould, even one made of wood and sodium hydroxide or sodium silicate or seawater (if commercially produced silicates are not available) is enough to fabric good quality block with this technique. The product is more cost efficient mainly because it uses only natural earth material available almost everywhere. The technology use neither preliminary transformation nor high temperatures ($T \leq 90$ °C).

The plant size is smaller because a group of ten workers can make blocks and bricks in a small area. It is much faster than all the technologies known because blocks can be made and use in a single day or in less than ten (10) hours.

The invention creates a revolution in the sense that is a new technique, which is both different and also similar to existing more energy intensive methods. It's the only technique that keeps the aesthetic aspect of basic raw materials.

Color of the starting materials is maintained. Colors range from brown through red. There are no patent or technical papers embodying any part of the invention or similar technology. The field is not heavily researched. Sample Prepreration and Processing We have made samples of the bricks using alkali activation and tested them (attached report) with different raw material from Senegal that proves that the concept is perfectly successful. Figure 1 depicts the performance of brick made from one soil material (Schlamm) from Senegal. These were mixed with 4, 8 or 12 M NaOH as a thick paste and cured at 120°C for the hours noted in the graph. It seems that 12 hours is enough to achieve final strength.

Because the values for 8 and 12 are the same, it appears that a very strong block could be made with 8M NaOH. A 4M block will cost ½ as much because it contains ½ the NaOH and still achieves a strength suitable for building.

PENNSTATE

INVENTION DISCLOSURE
Original Form Should Be Submitted To the Intellectual Property Office Via The
Department Research Dean or College at Least Two Weeks Prior to Public Disclosure.

1. Title of Invention:

LOW TEMPERATURE ALKALI ACTIVATED SILICATE BLOCK

2. Inventor Identification: (Attach separate sheets to accommodate more than six inventors).

Inventor Name	Inventor Title	Department Address	Home Address	E-mail Address	Work Phone	Citizenship
MOUHAMADOU BASSIR DIOP	ASSOCIATE PROFESSOR	MRL, 110 University Park PSU, PA 16802	415 S Atherton St #25 State College PA 16801	mbd130 @psu.edu	814 865 2434	SENEGAL
MICHAEL GRUTZECK	PROFESSOR	MRL, University Park PSU, PA 16802		gut@psu.edu		USA

3. Execution of Disclosure: This disclosure must be (1) signed and dated by all inventors, (2) read, understood, and signed by one technically qualified non-inventor witness, and (3) read and signed by the appropriate dean of research or administrative officer.

Inventor Signature:	Date: 2/15/06	Inventor Signature:	Date:
Inventor Signature:	Date: 2/15/06	Inventor Signature:	Date:
Inventor Signature:	Date:	Inventor Signature:	Date:

Witness Name: MARIA DICOLA

Signature of Witness: Date: 2/16/06

Name of Research Dean/Administrative Officer: CARLO G. PANTANO   Date: 4/20/06

Signature of Research Dean/Administrative Officer: CARLO G. PANTANO   Date: 3/20/06

NONCONFIDENTIAL DESCRIPTION

"*Low* Temperature Process to Create Bricks "

By M. B. Diop, et al.
PSU Invention Disclosure No. 2006-3160

Field of Invention

Construction, Residential, Industrial and Other; *Carbon Emissions Reduction*

Background

The manufacturing of bricks is an energy-intensive industry that contributes significantly to carbon dioxide emissions (http:www.epa.gov/ttn/chief/ap42/ ch11/final/c11sO₃.pdf). Emissions from brick manufacturing facilities include particulate matter (PM): PM less than or equal to 10 μm in aerodynamic diameter (PM-10) and PM less than or equal to 2.5 μm in aerodynamic diameter (PM-2.5), sulfur dioxide (SO_2), sulfur trioxide (SO_3), nitrogen oxide (NO_x) carbon monoxide (CO), carbon dioxide (CO_2), metals, total organic compounds (TOC), including methane, ethane, volatile organic compounds (VOC), and some hazardous air pollutants (HAC), hydrochloric acid HCl and fluoride compounds. More than 50 chemical pollutants are generated by the process.

In 2002, over eight billion bricks were sold in the United States. Typically, bricks are made by firing clay to approximately 2,000 ∘F. Considering the fact that the manufacturing of one brick consumes around 2 Mega Joules/kg brick, the total energy used in the U.S. can be estimated at sixteen billion Mega Joules/kg brick.

The heating process changes the molecular structure of the clay, such that it is vitrified. Variations in a brick's color, texture and performance characteristic may be made by changes in the mixture of clay, shale, water, air as well as any other special additives or coatings. Brick is the leading wall cladding material for the commercial market, while retaining a strong presence in the residential market, which consumes over eighty percent of all bricks made.

Description of Invention

The subject invention is a low temperature process to make bricks. The process occurs at near ambient temperatures and does not generate any pollutants (gases or particles) or waste. The process time lasts for several hours. The starting materials can consist of natural materials as well as by product materials. As such, *the inventors believe that* the process is more cost efficient than standard practices used in the industry. The characteristics of the brick covered by this invention include a controllable range of colors from red to brown. *Research results have shown that the ultimate compressive resistance of the invention's bricks range from 10 – 20 mPa, depending on processing conditions. This strength can be closely controlled by the process conditions (concentration, fineness, time of curing...etc.)* These results suggest a marked improvement over bricks currently on the marketplace. *Other characteristics, including water resistance, are comparable to existing bricks.* Leaching tests show that structure of the mineral that form during the process remain stable. Bricks generally have low solubility and projected durability. More testing is needed to verify these characteristics.

Invention Status

Invention has been reduced to practice and shall be utilized in the performance of a multiyear research project on durability at the Pennsylvania State University.

Utility

Bricks produced by this invention have construction applications, in the residential, industrial and governmental contruction markets.

For more information, contact: Matthew D. Smith, Intellectual Property Office, The Pennsylvania State University. 113 Technology Center, University Park, PA 16802, Phone: (814) 865-6277, Fax: (814) 865-3591, E-Mail: mds126@psu.edu

INDEX

K

K$^+$, 104
Kenya, 31
kinetics, 102, 108, 145, 147, 149
KOH, 160

L

leaching, 20, 68, 110, 123, 133, 143, 149, 153
limestone, 2, 7, 8, 18, 40, 43, 45, 53, 55, 56, 57, 58, 59, 73, 165
lithium, 16
lithography, 7, 59
low temperatures, 32, 53

M

macromolecules, 10
magnesium, 10, 16, 27, 43, 45, 164
manufacturing, vii, 2, 54, 55, 57, 73, 76, 90, 94, 105, 151, 168
manure, 68, 69
mass, 5, 43, 45, 46, 47, 49, 54, 66, 68, 70, 73, 88, 93, 140, 142, 159, 162
Material Research Laboratory, 2
matrix, 21, 22, 141, 143
matter, 2, 70, 71, 133, 168
Mauritania, 15
mechanical properties, 38, 75, 125, 126, 141, 143, 148, 150
melt, 152
melting, 46, 76
metals, 2, 27, 35, 150, 168
Mexico, 18, 66, 67
microcrystalline, 8
microscope, 98
microscopy, 27
microstructure, 39, 48, 50, 107, 117, 118, 128, 129, 130, 132, 135, 136, 137, 138, 143
Middle East, 63, 66
Miocene, 9, 20

moisture, 45, 70, 74, 95, 114, 164
mold, 52, 144
molecular structure, 2, 168
molybdenum, 16
mordenite, 32, 33, 36
MRI, 2, 157

N

Na$^+$, 37, 104, 143
Netherlands, 25
neutral, 10, 20, 22, 76, 138, 163
New Zealand, 32
nickel, 16
nitrates, 7, 153
nitrogen, 2, 25, 168
NMR, 37, 160, 163
non-polar, 160
North Africa, 66
North America, 51, 66, 67, 160
nucleation, 138

O

oceans, 6, 12, 143, 153
OH, 9, 10, 11, 28, 34, 35, 37, 49, 57, 97, 116, 118, 130, 132, 137, 138, 141, 145, 147
oil, 14, 18, 33, 71, 164
Oklahoma, 9
Oolitic limestones, 7
organic compounds, 2, 35, 56, 168
organic matter, 70, 71
oxidation, 12, 45
oxygen, 12, 30, 31, 34, 55, 152

P

Pakistan, 71
PCT, 110
Persian Gulf, 7
Peru, 40, 66, 68
pH, 6, 7, 10, 11, 13, 31, 32, 34, 35, 37, 50, 54, 59, 123, 133, 138, 139, 149, 153